T0190093

Environmental Footprints and Eco-design of Products and Processes

Series editor

Subramanian Senthilkannan Muthu, SgT Group and API, Hong Kong, Hong Kong

This series aims to broadly cover all the aspects related to environmental assessment of products, development of environmental and ecological indicators and eco-design of various products and processes. Below are the areas fall under the aims and scope of this series, but not limited to: Environmental Life Cycle Assessment; Social Life Cycle Assessment; Organizational and Product Carbon Footprints; Ecological, Energy and Water Footprints; Life cycle costing; Environmental and sustainable indicators; Environmental impact assessment methods and tools; Eco-design (sustainable design) aspects and tools; Biodegradation studies; Recycling; Solid waste management; Environmental and social audits; Green Purchasing and tools; Product environmental footprints; Environmental management standards and regulations; Eco-labels; Green Claims and green washing; Assessment of sustainability aspects.

More information about this series at http://www.springer.com/series/13340

Subramanian Senthilkannan Muthu
Editor

Development
and Quantification
of Sustainability Indicators

 Springer

Editor
Subramanian Senthilkannan Muthu
SgT Group and API
Hong Kong, Hong Kong

ISSN 2345-7651   ISSN 2345-766X (electronic)
Environmental Footprints and Eco-design of Products and Processes
ISBN 978-981-13-4790-0   ISBN 978-981-13-2556-4 (eBook)
https://doi.org/10.1007/978-981-13-2556-4

This Springer imprint is published by the registered company Springer Nature Singapore Pte Ltd.
The registered company address is: 152 Beach Road, #21-01/04 Gateway East, Singapore 189721, Singapore

This book is dedicated to:

*The lotus feet of my beloved Lord
Pazhaniandavar
My beloved late Father
My beloved Mother
My beloved Wife Karpagam
and Daughters—Anu and Karthika
My beloved Brother*

Contents

Sustainable City Indicators in Malaysia

Filzani Illia Ibrahim, Aisyah Abu Bakar and Dasimah Omar

Abstract At the district stages, sustainability is shaped by the place and the citizen's demand. As a result, the authorities resolve to the progress of sustainability pointers that quantify the progress of a community or a city. A large number of existing studies in the broader literature have examined the obstacles and approaches towards achieving sustainable cites. Among the discussed issues are the development of indicators that assess the level of sustainability of a city. Sustainable city indicators are the tool used to gauge the progress of a community or a city on selected themes or subjects relating to sustainability. Despite decades of research, the effectiveness of the sustainable city indicators is still debatable. There exist various types of sustainable indicators being studied and implemented nationally and internationally. This chapter focuses on sustainability indicators focusing in Malaysia. It is to analyse the notional context of indicators for sustainable city applied by the Malaysian authorities. The chapter also looks into the implementation of sustainable city indicators by the Shah Alam City Council as a case study. The gathered data are analysed using content analysis. The findings deliver an overall insight of sustainable city indicators used worldwide and the implementation of sustainable city indicators in Malaysia. This research facilitates the local authorities, professionals and urban planners in revising and improving the current use of sustainable city indicators.

F. I. Ibrahim (✉)
Faculty of Built Environment, Engineering Technology and Design,
School of Architecture Building and Design, Taylor's University,
Subang Jaya, Malaysia
e-mail: filzanillia@gmail.com

A. A. Bakar
Faculty of Environmental Studies, Department of Environmental Management,
Universiti Putra Malaysia, Seri Kembangan, Malaysia
e-mail: isya.ab@gmail.com

D. Omar
Centre of Town and Regional Planning, Universiti Teknologi Mara (UiTM),
Shah Alam, Malaysia
e-mail: dasimaho@yahoo.com

© Springer Nature Singapore Pte Ltd. 2019
S. S. Muthu (ed.), *Development and Quantification of Sustainability Indicators*,
Environmental Footprints and Eco-design of Products and Processes,
https://doi.org/10.1007/978-981-13-2556-4_1

Keywords Sustainable · Urban planning · Urban · Sustainability
Sustainable city · Indicators

1 Introduction

In general, sustainability is the ability to persevere. Subsequently, feasible advancement is an improvement that addresses the issues of the current issues without bargaining the requirements of what is to come. Hence, supportable improvement requires three primary compromises that are natural, social value and monetary maintainability. It gives a principle area to different maintainability guidelines as of late. Hence, reasonable advancement can be viewed as a method for creating urban zones by trading off both present and future ages' need. The diagram in Fig. 1.

The rise of sustainability thoughts amid the previous three decades has catalyzed large amounts of movement in creating 'actualities on maintainability' as markers. Numerous endeavours to create maintainability markers by analysts, and this incorporates Malaysian administration. Hence, this chapter will discuss about sustainable city in Malaysia.

Fig. 1 Sustainability domains

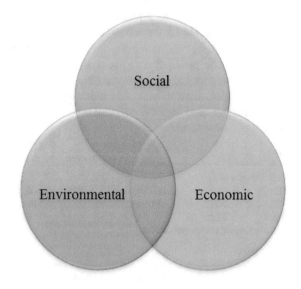

2 Social Indicator: The Movement

The development of social indicator investigation can be found as ahead of schedule as the late seventeenth century (Leiby 1970). Historical events identifying with the social marker development from seventeenth century to nineteenth century has given critical commitments on seeing social pointers are perceived in the present. The commitments exhibited are (i) inferential insights, (ii) pseudo-objectivity methodologies, (iii) inductive and deductive techniques.

According to Gasteyer and Flora (1999), Land and Ferriss (2006), the social pointer development had bit by bit motivated natural markers advancement from end of 1900s onwards. In the mid-1980, EIA and SIA were presented and progressively used (Gasteyer and Flora 1999; Land and Ferriss 2006). Thus, the effect evaluations were utilized for observing and envisioning the future outcomes of improvement activities on natural and social environment. Furthermore, the SIA worked as a component of EIA enactment which endeavoured towards envision and outcomes of extensive scale open ventures on social condition. Hence, the effect appraisal likewise utilized QoL and SWB measures to upgrade assessment of open-mediation endeavours and to screen key socio-mental state essential in deciphering social change (Gasteyer and Flora 1999; Land and Ferriss 2006; Sharpe 1999; Veenhoven 2002).

According to Cobb and Rixford (1998), Sharpe (1999), Valentin and Spangenberg (2000), Wu (2013), the idea of reasonable advancement, as featured by the Brundtland report and the Rio gathering, then transported another structure for social pointer development in the 1990s. New system for improvement was presented with solid accentuation on reasonable advancement.

Sustainability indicators then were utilized towards feature ramifications of present advancement drifts on what's to come. The structure joined calculated models showing how social advance, monetary development and natural concerns are interrelated (Cobb and Rixford 1998). In any case, in the 1990s, government offices or real national establishment demonstrated less backings in the supportability pointer advancement when contrasted with the most recent decades. Rather, the marker development from 1990s subsequently was all the more concentrating on network pointers (Cobb and Rixford 1998; Sharpe 1999).

Reasons for social indicator advancement either to understand or for down to earth activity to be made were investigated or tested over the long periods of social pointer development. Table 1 shows the four conflicting principles in social indicator development learned from the history of social indicator movement.

Table 1 Conflicting principles in social indicator movement

Conflicting principles	Explanations	
Nature of Indicator Statistics Descriptive statistics against Inferential statistics	Descriptive statistics used the data to provide descriptions of the sample either to numerical calculations, tables or graphs	Inferential statistics or the analytical approach used the data to make inferences, judgements and predictions
Data Collection Approach Pseudo-objectivity approach against Partisanship approach	The pseudo-objectivity approach or also known as the impartial or non-ideological process implies that the information gathered exist outside of the mind and in the real world. That is the data is not influenced by opinions or feelings. The assessment is based solely on the results of the experiment	Partisanship approach or also known as the ideological process implies that the data favours one side of ideas or feelings without considering other possibilities. The data adhered to a party, cause or person. The assessment is likely to result from prejudiced to nonrepresentational results
Analysis Approach Inductive methods against Deductive methods	Inductive method compile data about social conditions before making generalizations	Deductive method believes that social indicators should be based on abstract models that produce testable hypotheses
Purpose of Indicators Academic tools for understanding against rough guides for practical action	Academics anticipated that indicators should be kept from politics in order to obtain compiled data for several decades as well as to conduct pure research. Academics believed that conclusions published before time were premature	Practitioners needed to make judgment on model basis data and up-to-date information thus they were impatience for final answers

Source (Campbell et al. 1976; Cobb and Rixford 1998; Land and Ferriss 2006)

3 Conflicting Perspectives on Sustainability Indicators

Sustainability has all the earmarks of being barely alluded to natural upkeep (O'Brien 2012). The consideration set on condition comes since the root of the sustainability idea, which is the field of ranger service (Nations and Nigh 1980). Formerly, sustainability stresses that in no way, shape or form reaping ought to surpass what the backwoods yields or recovered in its next development (Kuhlman and Farrington 2010). Thus, sustainability is clarified towards a more extensive range; manageability is subjected to definitional ambiguity (Kuhlman and Farrington 2010; Kjell 2011). In the endeavour to abstain from abusing maintainability as a catchphrase compatible with 'goodness' in 'economical city', pragmatic elucidation of manageability is basic for the examination.

The most usually utilized definition among the three is by World Commission on Environment and Development. The definition has been additionally delineated into the three mainstays of supportability, specifically social, financial and natural (Kates et al. 2005; Strange and Bayley 2011). Some authors expressly bring up that the three columns or once in a while alluded to measurements ought to get equivalent weightage (Pope et al. 2004). Anyway others emphatically differ as they trusted that social and financial measurements worry on the present age, while natural measurement nurtures the future. That implies the weightage for the current age is twice as much as what's to come. The idea totally abused the Brundtland report saying that improvement ought not happen to the detriment without bounds age (Brundtland 1987). The three tended to definitions albeit prominent and direct, resist operational execution. The definitions welcome inquiries, for example, 'what is should be supported', 'to what degree of day and age' and 'what physical and social process appropriate in meeting the alluring points?' (Kuhlman and Farrington 2010; Kjell 2011).

The business analysts think about maintainability as a characteristic point in their field of study. Much of the time, the shortage of assets is the keyword in the horrid science. The effective meaning of sustainability over the financial expert viewpoint is the upkeep of aggregate capital esteem. Add up to capital esteem alludes to the total of fabricated capital, human capital, normal capital and social capital. Powerless supportability happens when characteristic capital is substitutable with different types of capital, though solid manageability denies the substitutability of regular capital, and consequently, the concentration movements to natural maintainability (Neumayer 2007; Dietz and Neumayer 2007; Kuhlman and Farrington 2010; Dedeurwaerdere 2013). The limit amongst solid and frail supportability is the degree of substitutability of common capital. Both solid and powerless supportability recommended that future age will acquire resources no not as much as the assets of the first era. Solid manageability suggests that the legacy is natural resources, while feeble maintainability infers that the legacy is riches comprising of man-made and ecological resources (Neumayer 2012; Kuhlman and Farrington 2010; Mancebo 2007; Dedeurwaerdere 2013).

The financial point of view gives more handy thoughtful and achievable processes of sustainability through the capital method. In any case, the capital method is imperilled to four exceptionally easy to refute requirements. The requirements are (i) nation particular information choice and accessibility (Booysen 2002; Böhringera and Jochemc 2007), (ii) the lawful framework overseeing the financial operator (Ayres et al. 1998; Kuhlman and Farrington 2010), (iii) flexibility of biological community and (iv) transboundary impacts.

Components of every capital might be hypothetically wide; hence, pointers are exposed to information accessibility in spite of the fact that information accessibility enhances later on. Selection of variables to constitute the capitals involves social and political inclination along these lines the choice and accessibility of factors can be nation particular (Booysen 2002; Böhringera and Jochemc 2007). Moreover, the monetary investigation on the substitutability is compelled by the standards and legitimate framework overseeing the financial operator

(Ayres et al. 1998; Kuhlman and Farrington 2010). Subsequently, the limit of substitutability of common capital likewise changes crosswise over nations. The limit is additionally very impacted by the versatility of the biological system to recuperate from stun or stretch. The biological system versatility decides the expendable assets and breaking points to the characteristic replenishable framework (Kuhlman and Farrington 2010).

Furthermore, it has turned into a developing worry in the globalized universe of how the exertion of a nation to accomplish supportability is influencing prosperity of neighbouring or different nations. The tended to imperatives to the capital method have been a warmed verbal confrontation since few decades prior (Kuhlman and Farrington 2010). Regardless, the capital approach is as yet honed broadly and universally up and coming (Dietz and Neumayer 2007; Neumayer 2012; Smajgl and Ward 2013; von Hauff 2016; Wilson and Wu 2017; de Albuquerque 2016).

While trying to reclassify Sustainable Development Goals embraced in by United Nations General Assembly, a few creators restrict with SDG strategies for top-down government controlling to oversee issues in maintainable improvement (Meuleman and Niestroy 2015; Hajer et al. 2015; Gupta et al. 2015). Ongoing studies tended to four transformative ways to deal with maintainability that ought to be worked with open investment. The primary approach involves ecological concerns and assurance of open merchandise in recognizing planetary limits. The second approach involves the significance of interconnected social and ecological worries and in addition issues of dispersion in featuring sheltered and simply working space. The third approach involves the readiness of a wide scope of performing artists to take activities in accomplishing enthusiastic culture. The last approach involves incitement of development and new practices through green rivalry (Hajer et al. 2015). Gatherings or network concerned ought to be locked in cooperatively in conveying numerous points of view to maintainable advancement because of their interests and limits (Crivits et al. 2016).

Hence, in estimating manageability, experts remain slanted towards target pointers, ideally quantitative information, and if vital, amiable for thorough insights. On the other hand, nearby networks will probably support subjective pointers, and if vital through subjective estimation, responsive to fundamental highlights of their conclusions (Fricker 1998; Morse et al. 2001). Three critical qualities in supportability talk are (i) quietude, (the deficiency of human learning); (ii) prudent, (alerts when in questions); and (iii) irreversibility, (decline irreversible alterations) (Viederman 1995). Many of the economic, social and environmental indicators aimed to measure sustainability levels fall short to provide an inclusive information (Fricker 1998).

Among indicators which are evident as inadequate in the literature include the ecological footprint, sustainability indicators, sustainable economic welfare indicators and quality of life indicators (Fricker 1998; Morse et al. 2001; Lehtonen et al. 2016). The greatest requirement featured is the human and social criteria. The pointers neglect to perceive the level of preparation and decisions individuals have in their activities. Other than specialized arrangements and strategy executions, supportability can basically be accomplished through people without decisions from any deliberate weight.

Regardless of vast assortment of writing on understanding estimating maintainability, the data conveyed are as yet sketchy. A few creators claimed the sensible method to quantify manageability is to isolate the necessities of the present from the requirements without bounds (Kuhlman and Farrington 2010). This thought trusted that the necessities of the present age should first be satisfied and later saving the assets for the people to come. However, gauging the needs of the present and the future beggar the minds on 'what is really needed?' and 'how much is enough?' Human needs, personal satisfaction and expectation for everyday comforts are not outright, as thoughts and inclinations change extra minutes (Fricker 1998; Kjell 2011). Accordingly, estimating manageability ought to be adaptable. In any case, maintainability is regularly appeared to be up to the experts and arrangement creators to take activities and steer people in general. The nationals' resourcefulness and capacity to adjust and take activities inside self-sustaining ecological points of confinement is prohibited from the condition. In humanism, manageability is a 'truant referent' (Fricker 1998). It calls for gatherings or supporters that are unreasonably overlooked. For this situation, the voting demographic is the neighbourhood network.

Sustainability frequently represents limits (Fricker 1998). Engineers of social pointers reports regularly tended to on the thoughts of breaking points in tending to manageability. Then again, supportable urban advancement is the procedure of mix and co-development among social, ecological and financial subsystems that make up a city (Kates et al. 2005). For better understanding, the part advances cases of universal social pointers that execute restrains and in addition the joining among social, ecological and financial subsystems.

4 Social Indicator Reports: The Examples

Among few models of indicator, two reports that are positively similar to Sustainable City Indicators in Malaysia are (i) Sustainable Society Index and (ii) Social Progress Index.

- Sustainable Society Index

Sustainable Society Index (SSI) in 2006 is to measure growth of societies towards reaching sustainability. It consists of 21 indicators whereby it have eight (8) categories and three (3) dimensions which are human, environmental well-being and economic.

The human well-being, indicators are categorized under (i) basic need, (ii) health and (iii) personal and social development. In environmental well-being, indicators are categorized under (iv) nature and environment, (v) natural resources and

(vi) climate and energy. In economic well-being, indicators are categorized under (vii) transition and (viii) economy. The three measurements are reliant. The measurement of human prosperity shows the basics of people. The natural prosperity speaks to the biological community where the people's live and monetary prosperity is the basics which people should have the capacity to do what they need.

SSF trusts that supportability is not just worry on consumption of assets; however, it remains on four standards. The first and the second standards are intra-generational value which alludes to solidarity in the present society and between generational value that is not to deny the earth and assets so that cutting edge would not live in insufficiency. The third is the ecological limits that is to live within Earth's carrying capacity. The fourth rule which SSF thought to be prudent is in case of deficient data; it is smarter to blunder in favour of alert as opposed to gambling irreversible decay. Accordingly, SSF perceived reasonable society as the general public that (i) addresses the issues of present age, (ii) does not trade off the assets of future age fundamental for their requirements and (iii) trusts that each individuals is given the chance to advance in opportunity inside very much adjusted society and in agreement with the environment (Mari and Jankovi 2014) (Table 2).

Seemingly, in spite of the fact that SSF underlined the significance of addressing needs and chances of present society without trading off the requirements for future age, the association of the pointers scarcely perceived cut-off points.

There are no sign of (i) fundamental needs of the general public which perceive components that people cannot live without, (ii) complimentary necessities of the general public which perceive components that would improve lives, yet without them, the existing framework is undisrupted and (iii) wanted open doors in life which may require alerts so accomplishing open doors would not bargain the assets without bounds age.

- The Social Progress Index

It is a sample of global social indicators that are demonstrated based on the needs. The three dimensions of SPI which are (i) Basic Human Needs, (ii) Foundation of Human Needs and (iii) Opportunity represent the report structure and components configuration of SPI.

The Social Progress Index (SPI) is created by an association called Social Progress Imperative. Social Progress Imperative is a non-benefit non-administrative association built up in 2012 in the USA. The SPI is the collected lists of social and ecological pointers. The files speak to the result of progress and not the degree of exertion that the nation makes. An illustration would be the accomplished well-being in a nation rather than how much the nation spent on well-being. In spite of the claim that SPI supplements monetary development, they do exclude markers of financial development, for example, GDP and work rate. The SPI skipped the greater part of the normal monetary markers to maintain a strategic distance from the use of financial intermediaries. The point of SPI is to solely quantify social advance through a mix of social and natural markers for given measurements.

Table 2 Sustainable society index

Dim.	Cat.	Indicators	Rationales	Indicator measurement	Sources
Human well-being	Basic needs	Sufficient food	Condition for the development of an individual	Number of undernourished people in % of total population	FAO 2006–2008
		Sufficient To drink	Condition for the development of an individual	Number of people as % of the total population, with sustainable access to an improved water source	WHO 2010—Unicef Joint Monitoring Programme
		Safe sanitation	Condition for the prevention and spreading of diseases that would severely hamper a person's development	Number of people in % of total population, with sustainable access to improved sanitation	WHO 2009 and UN Population Division 2009
	Health	Healthy life	Condition for development of each individual in a healthy way	Life expectancy at birth in number of healthy life years (HALE—Health Adjusted Life Expectancy)	WHO 2009 and UN Population Division 2009
		Clean air	Condition for human health	Air pollution in its effects on humans	EPI 2012
		Clean water	Condition for human health	Surface water quality	EPI 2010
	Personal and social development	Education	Condition for a full and balanced development of children	Combined gross enrolment ratio for primary, secondary and tertiary schools	UNESCO 2010
		Gender equality	Condition for a full and balanced development of all individuals and society at large	Gender Gap Index	World Economic Forum 2011
		Income distribution	Fair distribution of prosperity is a condition for sustainability	Ratio of income of the richest 10% to the poorest 10% of the people in a country	World Bank 2010
		Good governance	Condition for development of all people in freedom and harmony, within the framework of (international) rules and laws	The average of values of the six Governance	World Bank 2010

(continued)

Table 2 (continued)

Dim.	Cat.	Indicators	Rationales	Indicator measurement	Sources
Environmental well-being	Nature and environment	Air quality	Condition for ecological health	Air Pollution in its effects on nature	EPI 2012
		Biodiversity	Condition for perpetuating the function of nature, in all its aspects	Size of protected areas (in % of land area)	UNEP-WCMC 2010
	Natural resources	Renewable water resources	Measure of sustainable use of renewable water resources in order to prevent depletion of resources	Annual water withdrawals (m^3 per capita) as % of renewable water resources	Aquastat 2009
		Consumption	Measure of the use and depletion of material resources	Ecological Footprint minus Carbon Footprint	Global Footprint Network 2008
	Climate and energy	Renewable energy	Measure of sustainable use of renewable energy resources in order to prevent depletion of fossil resources and to reduce GHG emissions	Renewable energy as % of total energy consumption	IEA 2010
		Greenhouse gasses	Measure of main contribution to climate change, causing irreversible effects	CO^2 emissions per capita per year	IEA 2010
Economic well-being	Transition	Organic farming	Measure for progress of transition to sustainability	Area for organic farming in % of total agricultural area of a country	FiBL 2010
		Genuine savings	Measure for the true rate of savings, essential for sustainability	% of Gross National Income (GNI)	World Bank 2010
	Economy	Gross domestic product	(Inadequate) measure for (the growth of) the economy	GDP per capita, PPP, current international dollars	IMF 2011
		Employment	Access to the labour market is a condition for well-being for all people	Unemployment as % of total labour force	ILO, World Bank and CIA World Factbook 2011
		Public debt	Measure of a country's ability to make independent decisions with respect to budget allocation	The level of public debt of a country as % of GDP	IMF and CIA World Factbook

The markers in the social advance are composed under 12 segments, which are grouped under three measurements. Essential human needs allude to the rights to fundamental survival. This suggests having the capacity to live in safety, shield, adequate water nourishment, fundamental therapeutic attention to have the capacity to get by to development. The establishment of prosperity alludes to the ethical premise of satisfaction, which profoundly identifies with essential information, well-being and adjusted biological system. At long last, open door alludes to the uniformity of chance to all nationals. Uniformity of chance is likewise a key component in opportunity and freedom (Porter et al. 2015) (Table 3).

The model configuration of community pointers saw in SPI showed several phases of social advancement in which should be perceived in measuring the advance of a country. They are (i) fundamental requirements (the survival assets of the country); (ii) complimentary requirements, (general public ready to enhance and manage their lives); and (iii) wanted openings, (the nationals have the opportunity and flexibility to settle on their own decisions). The phases of social advancement give a superior concentration in satisfying necessities of the nationals. Satisfaction of essential or endurance requirements of the country empowers residents to endeavour to move their need from concentrating on physical satisfactions to concentrating on manageable employments.

SPI is an estimable case of social markers that are demonstrated in view of the chain of importance of necessities. In building up the social pointers, three important inquiries: (i) Does a nation accommodate its kin's most fundamental needs? (ii) Are the building hinders set up for people and networks to improve and maintain prosperity? (iii) Is there open door for all people to achieve their maximum capacity?. The three measurements of SPI that are (i) Basic Human Needs, (ii) Foundation of Human Needs and (iii) Opportunity speak to the parts and pointers of SPI (Porter et al. 2015).

The SPI coordinates markers crosswise over social parts for the structure of human needs. The parts are progressively sorted out in depicting need of social improvement. Hence, the pointers restricted to yield, result and effect markers.

Few similarities and differences are observed across SSI and SPI are summarized below:

- Report structure, particularly measurement and order profoundly, relies upon received hypothesis or establishment behind indictor advancement.
- All marker reports embrace the pseudo-objectivity approach of information accumulation. The speculations just refine which pointers to be utilized, yet the information is assembled crosswise over nations. The information accumulation process is not for specific gatherings to demonstrate any foreordained presumptions.
- The reports embrace inductive approach that incorporates information before making speculation for the populace.
- Data are estimated quantitatively and used as aides for viable activities.
- SSI utilizes target markers, and SPI utilizes both goal and subjective pointers.

Table 3 Social progress index 2014

D	Categories	Indicators	Sources
Basic human needs	Nutrition and basic medical care	Undernourishment	Food and Agriculture Organization of the U.N.
		Depth of food deficit	
		Maternal mortality rate	World Health Organization
		Stillbirth rate	
		Child mortality rate	U.N. Inter-agency Group for Child Mortality Estimation
		Deaths from infectious diseases	World Health Organization
	Water and sanitation	Access to piped water	World Health Organization/ UNICEF Joint Monitoring Programme for Water Supply and Sanitation
		Rural versus urban access to improved water source	
		Access to improved sanitation facilities	
	Shelter	Availability of affordable of housing	Gallup World Poll
		Access to electricity	U.N. Sustainable Energy for All Project
		Quality of electricity supply	World Economic Forum Global Competitiveness Report
		Indoor air pollution attributable deaths	Institute for Health Metrics and Evaluation
	Personal safety	Homicide rate	Institute for Economics and Peace
		Level of violent crime	
		Perceived criminality	
		Political terror	
		Traffic deaths	World Health Organization
Foundations of Well-being	Access to basic knowledge	Adult literacy rate	U.N. Educational, Scientific, and Cultural organization
		Primary school enrolment	
		Lower secondary school enrolment	
		Upper secondary school enrolment	
		Gender parity in secondary enrolment	
	Access to information and communications	Mobile telephone subscriptions	International Telecommunications Union
		Internet users	International Telecommunications Union
		Press Freedom Index	Reporters Without Borders
	Health and wellness	Life expectancy	World Development Indicators
			World Health Organization

(continued)

Table 3 (continued)

D	Categories	Indicators	Sources
		Non-communicable disease deaths between the ages of 30 and 70	
		Obesity rate	
		Outdoor air pollution attributable deaths	
		Suicide rate	Institute for Health Metrics and Evaluation
	Ecosystem sustainability	Greenhouse gas emissions	World Resources Institute
		Water withdrawals as a per cent of resources	
		Biodiversity and habitat	Environmental Performance Index using data from the World Database on Protected Areas maintained by the United Nations Environment Programme World Conservation Monitoring Centre
Opportunity	Personal rights	Political rights	Freedom House
		Freedom of speech	Cingranelli-Richards (CIRI) Human Rights Dataset
		Freedom of assembly/association	
		Freedom of movement	
		Private property rights	Heritage Foundation
	Personal freedom and choice	Freedom over life choices	Gallup World Poll
		Freedom of religion	Pew Research Center
		Modern slavery, human trafficking and child marriage	Walk Free Foundation's Global Slavery Index
		Satisfied demand for contraception	The Lancet
		Corruption	Transparency International
	Tolerance and inclusion	Women treated with respect	Gallup World Poll
		Tolerance for immigrants	
		Tolerance for homosexuals	
		Discrimination and violence against minorities	Fund for Peace's Failed State Index
		Religious tolerance	Pew Research Centre
		Community safety net	Gallup World Poll

(continued)

Table 3 (continued)

D	Categories	Indicators	Sources
	Access to advanced education	Years of tertiary schooling	Barro-Lee Educational Attainment Dataset
		Women's average years in school	Institute for Health Metrics and Evaluation
		Inequality in the attainment of education	United Nations Development Programme
		Number of globally ranked universities	Times Higher Education, QS World University Rankings, and Academic Ranking of World Universities

(*Source* Porter et al. 2015)

In view of the survey of the reports, the two reports underscore that GDP and monetary development cannot quantify prosperity or satisfaction of a country. It is likewise discovered that elucidation on hypothesis of establishment behind pointer improvement preceding marker determination helps with social affair precise markers to quantify and convey what should be estimated and conveyed.

5   Sustainable City Indicators

Urban communities and associations are concentrating on the improvement of pointers to quantify advance. A pointer gives data on the state or state of something. In terms of sustainability indicators, it can be define as a policy-relevant variable defined in such a way as to be measureable over time and space.

Supportability markers can be quantitative or subjective measures. Markers are frequently specified in strategy contemplates in the light of the fact that they claim to make accessible data about the linkages between various divisions and about developing patterns. Pointers are additionally crucial in the execution of the idea of maintainability in arranging and city improvement.

Pointers are most valuable in manageability arranging when connected to supportability limit or targets. Thresholds are scientifically determined points where the state of thing will change dramatically. Targets are regularly controlled by strategy creators or through open meeting and indicate level that must be met later on is manageability objectives are to be achieved.

Hence it can be stated that indicators used as a tool in monitoring and show the sustainability of city development and should closely associated with the main objective of sustainable development. Likewise, it can be characterized as to indicate slants and give quantitative and subjective data. Likewise, it goes about as one of the estimating instruments that can be utilized viably to empower invested individuals, to survey the accomplishment of a network, the network or a city. As per Mega (2010), supportability pointers and lists may fill in as compasses on the voyage to urban maintainability. They likewise go about as indispensable instruments for assessing, revealing and reorienting progress towards maintainable improvement.

The markers ought to have the capacity to demonstrate the level of supportability of a city. It likewise fills in as an apparatus for direction in supportability arrangements, including checking of measures and their outcomes and correspondence to the general population.

In any case, what separates them from commonplace natural or monetary markers is their attention on linkages crosswise over various divisions. In this way, to quantify the level of supportability of a city and the personal satisfaction of urban tenants, different urban pointers were produced seriously to give a sound premise to arranging and basic leadership to make a maintainable city worked. Urban supportability markers have been chosen as fundamental components for imparting the status of the training, which at that point serve to decide how fruitful procedures and strategies have been in the fulfilment of maintainability objectives.

6 Sustainable City Indicators: A Case Study of Malaysia

Malaysia struggles with low urban density but widely dispersed urban growth. It is prove that the fringe regions are becoming speedier to the detriment of the downtown areas. The lower cost of land at the provincial urban periphery and the urban edges, the extension of new thruways and the overwhelming dependence of private cars are among factors adding to the urban sprawl. The urban areas of the country are taking the form of automobile-oriented suburban growth (Jamalunlaili et al. 2009).

The present patterns of urban sprawl and nationals' ways of life are prompting the decrease in downtown areas, movement clog, wasteful utilization of land and the loss of green zones. Foundations, for example open transportation, are imperative to alleviate the issues. The inclination of the specialists to give careful consideration on financial and physical advancement contrasted with social improvement is reflected in the present condition of urban sprawl. Economic development should not isolate social progress in an attempt for urbanization. Change of social foundations should be expert at an indistinguishable pace from the populace development. Pivotal markers, for example condition, instruction, well-being and welfare, family and open support, ought to be dealt with as similarly as vital as the financial development in urbanization process.

Reasonable improvement has been a much-faced off regarding subject in the latest years particularly in creating nations. Because of quick urbanization with expanded populace and fast financial development, each creating nation is currently pushing ahead in actualizing maintainable improvement. The Malaysian

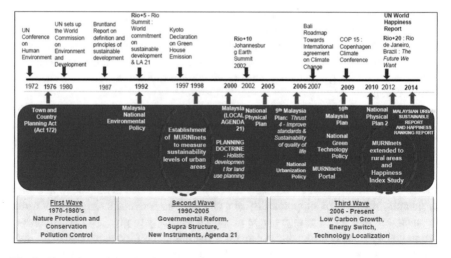

Fig. 2 Chronology of sustainable development in Malaysia

government has communicated its worry in accomplishing the objective of supportability through a few systems defined in their five-year national advancement
designs.

Hence, in Fig. 2 beneath demonstrates the outline real marker improvement and
foundation activity in Malaysia as far as objectives, principle highlights, phases of
execution, organizations securing these activities and the scales at which they work
including the maintainability pointers that will be talk about i.e. Malaysian
Urban-Rural-National Indicators Network on Sustainable Development
(MURNInets) (Table 4).

7 Malaysian Urban Rural National Indicators Network (MURNInet)

With regard to Malaysia, the quick development of populace and advancement has
encouraged the Federal Department of Town and Country Planning Peninsular
Malaysia, Ministry of Housing and Local Government Malaysia to create
MURNInets as a way to deal with measure and assess the maintainability of
Malaysia towns and urban communities.

Physical and spatial arranging under the setting of Malaysia is under the locale
of the Federal Town and Country Planning Department (FTCPD). Although its
mandate encompasses entire states it has most relevance in urban areas, and policies
it adopts can significantly influence the patterns of urban development in Malaysia.

Hence, pointers go about as a quantity outline of data about a subject or a
portrayal of the issue. It likewise makes an open door to the government in order to
provide more practical approach to challenges they are confronting and makes a

Table 4 Major indicator development initiative in Malaysia

	Goal	Main features	Implementation stage	Anchor agency	Scale
MQLI Malaysian Quality of Life Indicators	Expanding the measure of Malaysian success beyond economic achievement	A composite index showing the improvement in Malaysian Quality of Life with 1980 as the base year	Institutionalization	Macroeconomics and Evaluation Section of the Economic Planning Unit, Prime Minister's Department	National
Compendium of Environment Statistics (CES)	The integration of socio-economic information with environmental parameters	The statistics chosen are analysed according to the media based approach accommodating the Pressure-State-Response (PSR) model	Institutionalization	Department of Statistics (DOS)	National
Urban Sustainability Indicators and MURNInets (USI)	To design and test a set of urban indicators for the tracking of urban development towards sustainability	The first initiative in Malaysia linking indicators to benchmark values. MURNInet is the networked system, which will be used by local authorities to report on sustainability using the selected indicators	Testing	Federal Town and Country Planning Department (TCPD)	National
Malaysian Sustainable Development Indicators (MSDI)	Develop a national system for tracking Progress towards sustainability	Aiming to integrate sustainability elements into national level development planning	Identification	Environment and Natural Resource Section of the Economic Planning Unit, Prime Minister's Department	National
Klang Valley Regional Sustainable Quality of Life Index (KVRSQLI)	To develop stress ratio (spatial, growth and distributional weights) for the allocation of resources for the districts within Klang Valley	Index development Involving benchmarks at the district levels from a regional perspective	Formulation Completed	Federal Territory Development and Klang Valley Planning Division, Prime Minister's Department	Regional

(continued)

Table 4 (continued)

	Goal	Main features	Implementation stage	Anchor agency	Scale
Healthy City Indicators (HCI)	Continuously create social and physical environment for healthy urban population	Based on the World Health Organisation (WHO) framework. The community programme commenced in 1997 but its indicator development part is still at an early stage	Identification	Department of Health, Municipal Council of Kuching, Johor Bharu, Malacca	Local
Penang Report Card (PRC)	Define sustainable development for Penang utilizing a bottom-up participatory approach to planning	Based on the Sustainable Seattle model of active community-based Monitoring and organized by an NGO	Formulation Completed (one-off project)	Socio-economic and Environment Research Institute (SERI)	State
Sustainable Urban Development Indicators for the State of Selangor (SUDI)	Develop indicators to assess the improvement in urban issues such as water quality and waste management	The use of Environmental Management Systems (EMS) as the guiding framework	Identification	Sarawak Natural Resources Board	Local
Sustainable Development Indicators (SDIS)	Develop a state-level system for monitoring sustainability in cognisant of the state's administrative and legislative powers	Fitness-for-purpose indicator frameworks rather than the usual definitive suite of indicators	Identification	Town and Country Planning Department of Selangor	State

dream for the future they need to find in their city thinking about all parts of economy, condition and society.

The Malaysian Urban Indicators Network (MURNInets) application framework was made in view of a PC arrange intended to dissect exhibit urban conditions, impacts of improvement, to review transient changes and define practical urban situations for the future in light of settled gauges. The principle technique of MURNInets is to work towards improvement of the lives of city occupants by straightforwardly enhancing the base of data for participatory basic leadership for manageable human settlement advancement in Malaysia.

The framework fills in as an approach that can gauge the supportability of a city and locale through the 11 arranging areas. This approach is executed by every single Local Authority in Malaysia. There is an aggregate of 55 pointers utilized as a general sign of the maintainability of the city. The marker that is identified with this examination is in the part of open offices and diversion. The pointer is the proportion of aggregate land territory of open space to 1,000 individuals. Table 5 demonstrates the sequence of MURNInets usage programme.

Thus, several set of sustainability indicators are use, by using data as a key reference in the analysis of the sustainability of the area. The after-effect of examination will then be utilized by the neighbourhood specialist to centre around follow-up process for taking care of the issues and related issues in arranging and advancement. Through its implementation, a question raised in respect of the effectiveness of the system by using the indicators since it's launched several years ago.

Table 5 Chronology of MURNInets implementation programme

Year	Chronology
1998	Malaysian Urban Indicators Network (MURNInet) Design by FDTCP & Consultant 11 sectors 56 indicators
2002	Pilot Project on 6 cities
2004	Applied to 8 capital cities
2005–2006	Applied by all capital cities in Malaysia
2007–2009	MURNInet Portal LAUNCHED-online data entry 11 sectors 38 indicators
2010	MURNInet Indicator Modification Workshop 11 sectors 40 indicators
2011	Review and strengthening of MURNInet
2012	Malaysian Urban Rural National Indicators Network for Sustainable Development (MURNInets) 36 indicators 21 themes 6 dimensions MURNInets Portal Launched

Source Rashid (2015)

Table 6 MURNInets system

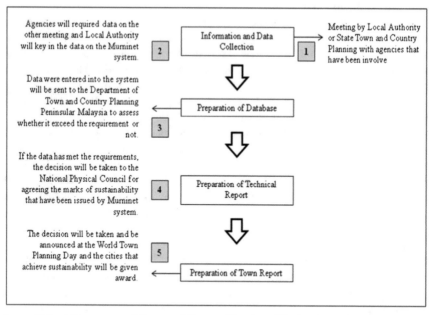

Source Federal Department of Town and Country Planning of Peninsular Malaysia (2004)

From the Federal Department of Town and Country Planning Peninsular Malaysia (2004), there are really five primary stages in the usage of MURNInets framework in Local Authority as shown in Table 6.

In any case, the framework confronted various issues. Thus, a few recommendations ought to be detailed to make another infusion for the MURNInet framework. It likewise incorporates some uniform pointers that are firmly identified with the assurance of maintainability criteria and kinds of information expected to meet the necessities.

8 Malaysia Happiness Index by MURNInet

The Happiness Index Study under the outline of the MURNInet comprises of background study and primary data collection. The HPI is organized in light of 6th push of the tenth Malaysia Plan: improving the standard and manageability of quality of life. That is, to create a caring society and promote community well-being.

71 out of 151 Local Authorities in Malaysia have participated in the nationwide survey. Bunch testing strategy is received, with 44,500 polls being conveyed all through Malaysia. The HPI is constituted under 13-single thing questions

Table 7 Happiness index questions

Foundations	Questions
Stress Level	How do you rank your stress level?
Health	Are you satisfied with your health?
Family Life	How satisfied are you with your life partner and family?
Job Satisfaction	Are you happy in your workplace?
Income	Are you satisfied with your current monthly income?
Spirituality	Do you believe that religion or spirituality can bring you joy and happiness?
Neighbourhood	Do you a good relationship with your neighbours?
Community	Do you enjoy it when you get involved with community activities in your area?
Safety	Do you feel safe in your home?
Public Amenities	Are you satisfied with the facilities provided in your neighbourhood?
Local Authority Services	Are you satisfied with services provided by local authorities in your area?
Political Representative	Are you satisfied with the service of the political representative in your area?
Environment	Are you happy with the quality of your living environment?

Source Rosly and Rashid 2013

estimating 13 establishments of bliss. The establishments are exemplified by Malaysia's Total Planning and Development Doctrine. The establishments incorporate feeling of anxiety, well-being, family life, work fulfilment, salary, other worldliness, neighbourhood, network, security, open comforts, nearby specialist administrations, political delegate and condition. A solitary inquiry speaks to every one of the angles. Table 7 demonstrates the 13-single thing poll.

The polls use the 5-point Likert scale; henceforth, the reaction choice change from 1 to 5. The satisfaction level extents from (i) 1.00 to 1.99, despondent; (ii) 2.00 to 2.99, less cheerful; (iii) 3.00 to 3.99, normal; (iv) 4.00 to 4.99, glad; to (v) 5.00, more joyful. The investigation approach utilized is graphic measurements. The bliss level is introduced through indicator investigation. The record is figured by separating the aggregate middle score with 65 (13 questions x 5 Likert scale). The item is the increased by 100.

Normal HPI for Malaysia is 76.06% (2014) contrasted with 74.80% (2013). The fundamental discoveries demonstrates the networks in nearby specialists that have scored high satisfaction positioning in accomplice and family, in profound accepts and in great associations with neighbours and network. MURNInets watch the perspectives with high scores as what matter most for the satisfaction of the networks. Likewise, the supporting situations for satisfaction are related to well-being, security, pay, happiness at the work put and proficient administrations and offices given by neighbourhood experts. Another intriguing discoveries uncover nearby experts in country settings have most elevated score in the joy positioning. The general population who live in the rustic regions are altogether more joyful with their lives than the general population who live in urban areas or any urban territories. The HPI is utilized as a device to figure related arrangements to enhance the social prosperity of networks.

9 (MURNInet): The Impact

As per federal government of Malaysia, an economical city is a city that can give the fundamental needs of city tenants, for example framework, metro luxuries, well-being and medicinal care, lodging, instruction, transportation, business, great administration and guarantee the populace needs, are met profiting all divisions of society

Subsequently, the effect of the urban maintainability markers can be seen from three points of view that are towards the elected and state government, neighbourhood specialist and nearby network and people in general. For the elected and state government, the markers can assess the city execution, enhance administrations for people in general, and can be viewed as fundamental assessment and estimation to update a urban region status and channel their speculation. Concerning neighbourhood specialist, it can recognize issues utilizing markers, urban quality issues and tending to those issues. Aside from that it can enhance benefit levels to the general population and to give criticism to National Integrity Plan. For the neighbourhood network and open, it can understand the exertion done by the administration that delicately mind towards people in general needs as far as administration level, offices level in the urban zone and nearby specialist.

Markers and their variations are not simply minor numbers. The instance of Malaysia, particularly MURNInets, is a unimportant case of an advancing institutional game plan arranging towards more fair and deductively tenable data to track urban manageability. The framework with its practical improvement markers for Malaysian towns and urban communities is relied upon to contribute towards demonstrating if the urban communities are indicating progress towards accomplishing manageable urban advancement destinations. Above all, it ought to give the truly necessary quality and precise learning and data at the city level for detailing powerful urban arrangements and projects towards satisfying the coveted urban destinations.

In this way, to have more strategy resounding urban pointers, more talk and research are expected to influence the framework to enhance every now and then. Henceforth, to improve the connections between manageability markers and approach process, a few definitions of systems ought to be utilized to advance the instrumental and theoretical utilization of pointers. Aside from that neighbourhood specialist included should look past information gathering and fundamentally analyse the issue of encompassing rather than just concentrating on the best way to accomplish a level of maintainability.

In any case, as there are as yet numerous issues that should be viewed as and better comprehended, the exertion done by the nearby expert for the execution of

MURNInets ought to be given credit. However, there are opportunities to get better and at last more practical entrance in implementation towards the approach frameworks.

Acknowledgements This chapter is dedicated to Taylors University, UPM and Universiti Teknologi MARA. Appreciation to the co-researcher in the completion of this chapter.

References

Ayres, R. U., Jeroen Van Den Bergh C. J. M. & Gowdy, J. M. (1998). Viewpoint: Weak versus strong sustainability. In *Tinbergen Institute Discussion Papers* (pp. 98–103). Retrieved from http://www.tinbergen.nl.

Böhringera, C., & Jochemc, P. E. P. (2007). Measuring the immeasurable—A survey ofsustainability indices. *Ecological Economics, 63*(1), 1–8.

Booysen, F. (2002). An overview and evaluation of composite indices of development. *SocialIndicators Research, 59*(2), 115–151. https://doi.org/10.1023/A:1016275505152

Brundtland, G. H. (1987). World commission on environment and development. *Our Common Future.* https://doi.org/10.1080/07488008808408783

Campbell, A., Converse, P. E. P. E., & Rodgers, W. L. W. L. (1976). *The Quality of AmericanLife: Perceptions, Evaluations and Satisfaction.* New York: Russell Sage Foundation.

Cobb, C. W., & Rixford, C. (1998). *Lessons learned from the history of social indicators: Redefining progress.* San Francisco: Redefining Progress. https://doi.org/10.2307/492843.

Crivits, M., Prové, C., Block, T., & Dessein, J. (2016). Four Perspectives of Sustainability Appliedto the Local Food Strategy of Ghent (Belgium): Need for a Cycle of Democratic Participation?*Sustainability, 8*(1), 55. https://doi.org/10.3390/su8010055

de Albuquerque, M. F. C. (2016). The sustainable use of biodiversity and its implications in agriculture: The agroforestry case in the brazilian legal framework. In *Legal Aspects of Sustainable Development* (pp. 585–606). Cham: Springer International Publishing. https://doi.org/10.1007/978-3-319-26021-1_29.

Dedeurwaerdere, T. (2013). *Sustainability Science for Strong Sustainability.* Université Catholique de Louvain.

Dietz, S., & Neumayer, E. (2007). Weak and strong sustainability in the SEEA: Concepts andmeasurement. *Ecological Economics, 61*(4), 617–626. https://doi.org/10.1016/j.ecolecon.2006.09.007

Fricker, A. (1998). Measuring up to sustainability. *Futures 30*(4):367–375. https://doi.org/10.1016/S0016-3287(98)00041-X

Gasteyer, B. S., & Flora, C. B. (1999). *Social Indicators : An Annotated Bibliography on Trends, Sources and Development, 1960–1998.*

Gupta, J., Pfeffer, K., Ros-Tonen, M., & Verrest, H. (2015). An inclusive development perspective on the geographies of urban governance. In *Geographies of Urban Governance* (pp. 217–228). Cham: Springer International Publishing. https://doi.org/10.1007/978-3-319-21272-2_11.

Hajer, M., Nilsson, M., Raworth, K., Bakker, P., Berkhout, F., de Boer, Y. etc. (2015). Beyond cockpit-ism: Four insights to enhance the transformative potential of the sustainable development goals. *Sustainability (Switzerland), 7*(2), 1651–1660. https://doi.org/10.3390/su7021651.

Jamalunlaili, A., Zulhafidz, M. Y., Mohd Zuwairi, M. Y., & Mohd Shakir, M. A. S. (2009). UrbanSprawl in Malaysia. *Journal of the Malaysian Institute of Planners,7*:69–82

Kates, R. W., Parris, T. M., & Leiserowitz, A. A. (2005). What is sustainable development? Goals, indicators, values, and practice. *Environment: Science and Policy, 47*(3), 8–21. https://doi.org/10.1080/00139157.2005.10524444.

Kjell, O. N. E. (2011). Sustainable well-being: A potential synergy between sustainability andwell-being research. *Review of General Psychology, 15*(3):255–266. https://doi.org/10.1037/a0024603

Kuhlman, T., & Farrington, J. (2010). What is Sustainability? *Sustainability, 2*(11):3436–3448. https://doi.org/10.3390/su2113436

Land, K. C., & Ferriss, A. L. (2006). The sociology of social indicators. In *21st century sociology: A reference handbook*. (pp. 518–526). United Kingdom: Sage Publications.

Lehtonen, M., Sébastien, L., & Bauler, T. (2016). The multiple roles of sustainability indicators ininformational governance: Between intended use and unanticipated inffiuence. *Current Opinionin Environmental Sustainability*. https://doi.org/10.1016/j.cosust.2015.05.009

Leiby, J. (1970). *Carroll Wright and Labor Reform: The Origin of Labor Statistics—James Leiby—Google Books.* Cambridge, MA: Harvard University Press

Mancebo, F. (2007). Le développement durable en question(s). *CyberGeo, 2007.* https://doi.org/10.4000/cybergeo.10913.

Mari, S. M., & Jankovi, S. M. (2014). Towards a Framework for Evaluating Sustainable SocietyIndex. *Romanian Statistical Review, 62*(3), 49–62.

Meuleman, L., & Niestroy, I. (2015). Common but differentiated governance: A metagovernance approach to make the SDGs work. *Sustainability (Switzerland), 7*(9):12295–12321. https://doi.org/10.3390/su70912295

Morse, S., McNamara, N., Acholo, M., & Okwoli, B. (2001). Sustainability indicators: The problem of integration. *Sustainable Development, 9*(1):1–15. https://doi.org/10.1002/sd.148

Nations, J., & Nigh, R. B. (1980). The evolutionary potential of lacandon Maya sustained-yield tropical forest agriculture. *Journal of Anthropological Research, 36*(1), 1–30.

Neumayer, E. (2007). Sustainability and well-being. In M. McGillivray (Ed.), *Human well-being: Concept and measurement* (pp. 193–213). Palgrave Macmillan UK: United Nations University. https://doi.org/10.1057/9780230625600_8.

Neumayer, E. (2012). Human development and sustainability. *Journal of Human Development and Capabilities, 13*(4):561–579. https://doi.org/10.1080/19452829.2012.693067

O'Brien, C. (2012). Sustainable happiness and well-being: Future directions for positive psychology. *Psychology, 3*(12):1196–1201. https://doi.org/10.4236/psych.2012.312A177

Pope, J., Annandale, D., & Morrison-Saunders, A. (2004). Conceptualising sustainabilityassessment. *Environmental Impact Assessment Review.* https://doi.org/10.1016/j.eiar.2004.03.001

Porter, M. E., Stern, S., & Artavia Loría, R. (2015). *Social Progress Index, 2015.*

Rosly, D., & Rashid, A. A. (2013). Incorporating happiness and well-being into sustainable development: Indicators framework. *Malaysian Townplan Journal, 9*(1).

Sharpe, A. (1999). *A survey of indicators of economic and social well-being.* Ottawa: Canadian Policy Research Networks, Centre for the Study of Living Standards

Smajgl, A., & Ward, J. (2013). *The water-food-energy nexus in the Mekong region: Assessing development strategies considering cross-sectoral and transboundary impacts* (Vol. 9781461461). https://doi.org/10.1007/978-1-4614-6120-3.

Strange, T., & Bayley, A. (2011) Sustainable development—linking economy, society, environment. *Energy Policy, 39*(10–2):6082–6099. https://doi.org/10.1016/j.enpol.2011.07.009

Valentin, A., & Spangenberg, J. H. (2000). A guide to community sustainability indicators. *Environmental Impact Assessment Review, 20*(3), 381–392. https://doi.org/10.1016/S0195-9255(00)00049-4

Veenhoven, R. (2002). Why social policy needs subjective indicators. *Social Indicators Research, 58*(1–3):33–45

Viederman, S. (1995). Knowledge for sustainable development: what do we need to know? In T. Trzyna (Ed.). *A sustainable world: Defining and measuring sustainability* (pp. 37–40). Sacra-Mentoo: IUCN.

von Hauff, M. (2016). Sustainable development in economics. In *Sustainability science* (pp. 99–107). Netherlands: Springer.

Wilson, M. C., & Wu, J. (2017). The problems of weak sustainability and associated indicators. *International Journal of Sustainable Development and World Ecology, 24*(1):44–51. https://doi.org/10.1080/13504509.2015.1136360

Wu, J. (2013). Landscape sustainability science: Ecosystem services and human well-being inchanging landscapes. *Landscape Ecology, 28*(6):999–1023. https://doi.org/10.1007/s10980-013-9894-9

A Mining Industry Sustainability Index: Experiences from Gold and Uranium Sectors

Issaka Dialga

Abstract From a sustainable development perspective, companies have to incorporate new requirements into their business models. Taking into account, new societal challenges are conceptualized at the corporate level by the concept of Corporate Social Responsibility (CSR). At the level of the mining industries, the expectations are much more important especially since the mining activity generates social impacts (creation of employment, but also prostitutions, child labor, precarious working conditions in the artisanal exploitations), environmental (pollution, noise pollution, loss of biodiversity) and economic ones (income increase, dynamism of the local economy, but the mining activity creates economic distortions such as the increase in the price of real estate, conflicts of land use). The net impact of mining activity is therefore sometimes difficult to measure. The tool most commonly used by companies subject to the CSR requirement is the Global Reporting Initiative. This standard tool cannot, however, account for the specificities of the mining sector or the singularity of the contexts. This chapter focuses on mining sector as a driver of local development by analyzing its contributions to key dimensions of sustainable mining sector. For needs of policy decision making, we suggest a composite Mining Industry Sustainability Index (MISI). The sensitivity and robustness analysis and the correlation tests with other well-known indicators, at the end of the chapter, prove the strength of the constructed index, namely the sustainability index of the mining industry.

Keywords Top-down and bottom-up approach · Mining industry
Gold · Uranium · Local sustainable development · Local communities' well-being

I. Dialga (✉)
LEMNA, Laboratory of Economics and Management of Nantes-Atlantique,
Department of Economics, University of Nantes, Nantes, France
e-mail: issaka.dialga@univ-nantes.fr

I. Dialga
LAPE, Laboratory of Economic Policy Analysis, Department of Economics,
University Ouaga II, Ouagadougou, Burkina Faso

I. Dialga
Chemin de la Censive du Tertre, 44322 Nantes Cedex 3, France

© Springer Nature Singapore Pte Ltd. 2019
S. S. Muthu (ed.), *Development and Quantification of Sustainability Indicators*,
Environmental Footprints and Eco-design of Products and Processes,
https://doi.org/10.1007/978-981-13-2556-4_2

Table of key abbreviations

BAP	Budget Allocation Process
BEEEI-NIGER	Bureau d'Evaluation Environnementale et des Etudes d'Impact
BUNEE-BF	Bureau National des Evaluations Environnementales
CMB	Chambre des Mines du Burkina
CBI	Children Well-Being Indicator
CCI	Control of Corruption Indicator
CI	Composite Index/Indicator
CPA	Principal Components Analysis
CPI	Cleaner Production Index
CSR	Corporate Social Responsibility
CWI	Communities' Well-Being Index
GRI	Global Reporting Initiative
GTI	Governance and Transparency Index
IEcoI	Industrial Ecology Indicator
IEI	Intra-Generational Equity Indicator
IGEI	Intergenerational Equity Indicator
LDI	Local Development Index
LEI	Local Employability Indicator
LSD	Local Sustainable Development
MEI	Mining Sector Employability Indicator
MGI	Mining Sector Governance Indicator
MISI	Mining Industry Sustainability Index
MILDI	Mining Industry's Contribution to Local Development Indicator
NEI	Net Environmental Impact Indicator
NR	Natural Resources
NRGI	Natural Resource Governance Institute
PPI	Purchasing Power Indicator
RAJIT	Réseau Africain des Journalistes pour l'Intégrité et la Transparence
RDI	Rural Development Indicator
SEI	Structuring Effects Index
TTI	Technology Transfer Indicator
WDI	World Development Indicators

1 Introduction

Sustainable development concept (see Brundtland 1987) has created new requirements. Indeed, at the Earth Summit held in Johannesburg in 2002 where large
companies were represented, the latter saw the emergence of a new demand
addressed to them. It is a question of evaluating the impacts of the activities of the

companies on their global environment, starting with the impacts on the personnel of the company and in a widened way to all the partners of the structure (customers, suppliers, sub-contractors, etc.) and the natural environment. From then on, the constraints faced by a company go beyond the simple economic profitability. Companies must take into account the following issues: governance of their structure, human rights in the enterprise, conditions and working relationships, environmental responsibility, the loyalty of practices, consumer and consumer protection issues, and communities and local development integration in the business model of the company.

Companies have to incorporate these new requirements into their business models. Taking into account new societal challenges is conceptualized at the corporate level by the concept of Corporate Social Responsibility (CSR). At the level of the mining industries, the expectations are much more important especially since the mining activity generates social impacts (creation of employment, but also prostitutions, child labor, precarious working conditions in the artisanal exploitations), environmental (pollution, noise pollution, loss of biodiversity) and economic ones (income increase, dynamism of the local economy, but the mining activity also creates economic distortions such as the increase in the price of real estate, conflicts of land use). The net impact of mining activity is therefore sometimes difficult to measure. The tool most commonly used by companies subject to the CSR requirement is the Global Reporting Initiative or the ISO 26 000 standard. These standard tools cannot, however, account for the specificities of the mining sector or the singularity of the contexts.

In a previous work (see Dialga 2018), we challenged to take into account the mentioned mining sector specificities and countries contexts. We analyzed how does mining industry affect the level of sustainable development across a country? In this chapter, we focus on mining sector as a driver of local development by analyzing its contributions to key dimensions of a sustainable mining sector. In other words, the aim of this chapter is to show how does the mining sector contribute to the expansion of other economic sectors or some components of sustainable development. The chapter seeks also to measure this sustainability at the mining industry level through a Mining Industry Sustainability Index (MISI). The MISI is built for needs of policy decision making. We analyze both dimensional scores and the composite score of the MISI using data from Burkina Faso and Niger mining industries. The methodological approach adopted in this study is the so-called *top- down/bottom-up* approach. The chapter is organized as follows. Section 2 describes the *top-down/bottom-up* approach used to build the MISI. Section 3 establishes the theoretical framework of the index. This section identifies the key components of the MISI in the case of Gold and Uranium industries. We also define indicators for each component of the index. Section 4 gives details on data collection and the sources used for the collection of such data. Section 5 justifies the choice of normalization, weighting, and aggregation of the MISI. Sect. 6 analyzes the robustness and the validity of the MISI. Section 7 concludes by summarizing the highlights of the study.

2   Methodology

The construction of the MISI is based on the so-called *top-down bottom-up hybrid* approach. This hybrid approach mobilizes both academic knowledge, professional experiences and life experience of citizens.

2.1   The Top-Down Approach

The top-down approach is mainly used in economics or environmental sciences. It is justified by the fact that unlike the objects of study of the social sciences (Sociology, Psychology), those of disciplines such as Economics or Environment are "silent" and the expert is responsible for thinking, conceiving and deciding of what is best. Indices are therefore developed from international initiatives set up as standards or through experts' works based on theoretical formulations that apply to all entities of the study without differentiation. The top-down approach is an inductive approach. It allows for a homogenization of practices. This approach has the advantage of simplifying the reality we want to measure. As the top-down approach is based on a methodical and rigorous approach, it gives the constructed indices a scientific legitimacy as the bottom-up approach confers them a local legitimacy. One of the reasons that also motivate the use of the top-down approach is the possibility of transferability of the results obtained by dissemination and the availability of data in international databases for monitoring indices over time. Indeed, the knowledge of the issue treated by the experts thanks to the accumulation of experiences allows them to have perspective on the topic and to treat it objectively (Reed et al. 2006) while specific and local experiments (bottom-up approach) present the risk of producing highly contextualized tools. The indices obtained by this approach are sometimes stronger compared to those derived from local experiences.

2.2   The Bottom-Up Approach

Reed et al. (2006), Chamaret (2007), O'Connor and Spangenberg (2008) argue that the bottom-up approach responds to a crucial need for information that responds to social demand and the specificities of contexts. It is anchored on an approach aiming at revealing the points of view, sometimes divergent, and issues specific to each context. Unlike the top-down approach described above, the bottom-up approach does not establish itself to the stakeholders of the study. The bottom-up approach allows to initiate a tool which will be the result of a collective construction and thus legitimized by a large public concerned in the process. By involving the stakeholders, the approach makes it possible to take into account the real knowledge of the parties concerned with the issues. Some of these concerns are sometimes ignored by the experts.

One of the advantages of the participatory approach is the possibility of generating qualitative information in order to "lubricate" the quantitative index. It is also possible, by this method, to obtain a representative sample by means of a random draw. This gives a scientific character to the approach.

However, the approach may be ineffective given the different levels of issues to be taken into account and the diversity of stakeholders who have interests that are often divergent. The question that arises is how to organize the stakeholders in the discussions and how to benefit from their diverse contributions?

2.3   The Top-Down and Bottom-Up Hybrid Approach

Top-down and bottom-up approaches are traditionally opposed. As described above, the top-down approach, also called "expert approach", goes from the top, either from international initiatives, or through nationally established standards or from the academic literature, to develop the tool which will have to be applied at a lower scale (at the local level, for example). While the bottom-up approach is an upward movement that goes from the lower echelon to the level of decision-makers.

Case studies have shown the relevance of hybrid approaches. Indeed, it is proven that in the presence of issues involving multidimensional concepts, such as the concept of sustainable development at the scale of the mining industry, and a variety of stakeholders (local government, local communities, unions, NGOs and mining companies) with the most often conflicting interests, it is increasingly appropriate to adopt a hybrid approach (Chamaret 2007, O'Connor and Spangenberg 2008). In addition, the multiplicity of issues at the social level (equity and governance), economic (profitability and performance) and environmental (nuisance and environmental damage) makes the use of such approaches relevant.

Top-down and bottom-up approaches have limitations when they are used separately, but combining the two approaches is an original approach to overcome the individual limitations identified in the literature.

Table 1 shows that the two approaches proceed in the same way, the only difference being that their angle of attack justifies the relevance of using a combination of the two approaches. Indeed, while in the composite index definition step, the top-down approach uses coherent theoretical frameworks or pre-established standards, the bottom-up approach involves community consultation, defines the most relevant variables based on their experiences and daily experiences. Similarly, in the data collection step, for example, the approach is left to the community initiatives that provide both qualitative and quantitative data for their own use and for the purposes of other stakeholders. This task is carried out by the expert in the top-down approach, which can only use existing data, based on international databases and essentially consisting of quantitative data.

This hybrid approach is more and more implemented in the field of composite indicators (see Rebai et al. 2016; Caiado et al. 2017; Gandhi et al. 2018).

Table 1 Two methodological paradigms for developing and applying sustainability indicators at local scales and how each method approaches four basic steps

Methodological paradigm	Step 1: Establish context	Step 2: Establish sustainability goals and strategies	Step 3: Identify, evaluate and select indicators	Step 4: Collect data to monitor progress
Top-down	Typically land use or environmental system boundaries define the context in which indicators are developed, such as a watershed or agricultural system	Natural scientists identify key ecological conditions that they fell must be maintained to ensure system integrity	Based on expert knowledge, researchers identify indicators that are widely accepted in the scientific community and select the most appropriate indicators using a list of pre-set evaluation criteria	Indicators are used by experts to collect quantitative data which they analyse to monitor environmental change
Bottom-up	Context is established through local community consultation that identifies strengths, weakness, opportunities and threats for specific systems	Multi-stakeholder process identify sometimes competing visions, end-state goals and scenarios for sustainability	Communities identify potential indicators, evaluate them against their own (potentially weighted) criteria and select indicators they can use	Indicators are used by communities to collect quantitative or qualitative data that they can analyse to monitor progress toward their sustainability goals

Source Reed et al. (2006)

Applying the top-down and bottom-up hybrid approach to our MISI's construction requires the following steps: select a set of candidate[1] indicators. The relevance of the selected indicators is based on literature, standards and case studies on mining industry.[2] Using a survey form, we then ask respondents to rank the

---

[1]The selected indicators are considered as candidate indicators because after the field survey, some of them could be dropped. It is also possible to integrate new indicators.

[2]Reader can find more details on mining case studies in Carney (1998), Hausmann and Rigobon (2003), Hugon (2009), Hilson (2010), Andersen and Aslaksen (2008), Van der Ploeg (2011), Arezki et al. (2012), Jensen and Wantchekon (2004), Auty (2001), Chen et al. (2010), Papyrakis and Gerlagh (2004), Bolt et al. (2005), Sba-Ecosys-Cedres (2011), Reed et al. (2006), Kaufmann et al. (2009), Brollo et al. (2010), Frankel (2010), Oxfam (2013), Geiregat and Yang (2013), Bardhan (1997), Dialga (2015).

candidate indicators following the importance/relevance of each indicator in mining industry sustainability issues. Our participatory survey involved four categories of development stakeholders: local government (municipal councils), mining companies, NGOs, and citizens. We also involved technical and specialized service structures such as the Ministry of Environment and green growth, the Ministry of Economy, the Ministry of Mines, Petroleum and careers, the Department of Child Protection, the Ministries of Agriculture and Livestock, the Directions of statistical studies, the Directions of studies and planning, as well as the Design bureaus and environmental impact assessment. The mining chambers represented the interests of the mining companies. Civil society organizations are also involved in order to take into account the issues of human rights, the environment, and local development. Citizens and local residents close to mining exploitation are also involved. In order to reduce the bias, respondents are randomly chosen. We also introduced three open questions in the questionnaire. The questions submitted to our 76 respondents focused, on the one hand, on the respondents' conception of what a responsible mining company is and, on the other hand, their expectations in terms of local development that would directly impact their community. Responses to these questions intended to complete the list of issues identified in the literature.[3] The ranking of candidate indicators by the respondents allowed us to generate a weight for each individual indicator through the so-called the Budget Allocation Process (BAP). Details on the BAP approach is described in Sect. 4. Finally, we analyze the results by looking at both composite sustainable mining scores and performance in each component of the index. We define a sustainable mining industry rule in order to classify the two mining sectors (Gold and Uranium): On a scale of 100, or considering the core components of this index, mining sector which obtains more than 67 of composite score or which performs more than 50% in 2/3 of the five components of the index is considered as sustainable. A lexical analysis (qualitative analysis) is also performed to better understand the composite scores.

3　Theoretical Framework, Dimensions, and Sub-indicators Identification

3.1　Theoretical Framework

The issue of sustainable development in relation to natural resources has led to two opposing visions. On the one hand, we have the strong sustainability defended by

---

[3]This literature includes Adiansyah et al. (2015), Rebai et al. (2016), Bai et al. (2017), Marin et al. (2016), Salisu Barau et al. (2016), Wang et al. (2018), Heravi et al. (2015), Kalsoom and Khanam (2017), Bini et al. (2018), Lu et al. (2017), Gastauer et al. (2018), Hodge (2014), Piccinno et al. (2018), Gandhi et al. (2018), Zhang et al. (2018), Hallstedt (2017), Valenzuela-Venegas et al. (2016), Kanyimba et al. (2014), Lagos et al. (2018), Kitula (2006), Caiado et al. (2017), Tahmasebi et al. (2018), Graetz (2014) and Sullivan et al. (2018), Northey et al. (2016).

authors like Daly (1990), Costanza (1992), Pearce et al. (1993) and on the other hand, there is the weak sustainability whose theoretical foundations come from the works of Hotelling (1931) and Hartwick (1977, 1990). Strong sustainability requires the maintenance of the entire natural resource[4] (see Daly 1990, p. 4, Costanza 1992, p. 16, Pearce et al. 1993, pp. 52–53) while weak sustainability admits a deterioration of the level of the stock of natural capital exploited in return for a reinvestment of the rents resulting from its extraction in capital reproducible such as machines, skills, health, governance, and institutions. For proponents of strong sustainability, Nature—and all that makes it up—is irreplaceable, and every element of Nature is also indispensable to the achievement of the well-being on Earth (Daly 1990). Unlike renewable natural resources (fish, forest) which provide several vital functions at the same time (maintaining the eco-systemic equilibrium, carbon sequestration, food source, etc.), non-renewable natural resources have very few eco-systemic values. Maintaining their stock is not an end in itself. This is even a utopia for developing countries to not extract these non-renewable natural resources. Indeed, countries rich in natural resources run a huge risk to lose their comparative advantage when they choose a strong vision of sustainable development (Gelb 2010).

The weak sustainability approach seems to us to be more justifiable than the strong sustainability approach, given the inevitably exhaustible nature of gold and uranium that are discussed in this chapter. The construction of the MISI is based on the theoretical framework of the Hartwick rule (1977) defined above.

The MISI's theoretical framework is illustrated by Fig. 1. We assume a perfect one-way substitutability between the mining industry and the five key components of a Local Sustainable Development (LSD). The main idea underlying this hypothesis is that the mining sector should have spillover effects for the other sectors of the national economy and particularly on local economy. The perfect substitutability is due to the fact that the reinvestment of the entire mining revenue (Hartwick 1977 rule) improves the performance of the five pillars of development. As highlighted in Yakovleva et al. (2017), mining industry can maximize its contribution to the UN Sustainable Development Goals in sub-Saharan Africa. Mining companies could further collaborate with other sectors to support the provision of collective goods. However, it is not possible to reconstitute the mining resource stock after its depletion. We postulate that the indicators within the same dimension may be substitutable, i.e., it is possible to compensate the loss of an indicator by improvement of another indicator. By contrast, the substitutability is imperfect between dimensions/pillars. That is because each pillar reflects a different important aspect of the mining industry and the LSD issues.

---

[4]"*Sustainable development requires that natural capital be maintained intact*" (Daly 1990, p. 4). Costanza uses the term "*constancy of total natural capital*" (Costanza 1992, p. 16).

3.2 The MISI's Dimensions and Sub-indicators

According to our knowledge on local sustainable development with mining sector as a main key driver, case studies[5] on mining issues and experiences of stakeholders across the field survey, five key dimensions of the sustainability of mining sector[6] should be taken into account. The five key components (see Fig. 1) and their sub-indicators are described below.

3.2.1 Local Development: Mining Industry's Performance and Its Spillover Effects on the Local Economy

This component of the index seeks to take into account the contribution of mining industry to local development through the spillover effects of the sector (stimulation of local demand upstream, creation of jobs for specific and vulnerable groups (women and youth), diversification of sources of income and increase of income. The improvement of the Local communities' well-being is measured by the improvement of social infrastructures such as transport infrastructure (roads), sanitation infrastructure, schools and trades training facility centers, sports and leisure centers, health centers, access to drinking water. In order to make tangible the various contributions of the mining sector in this component, we define the following indicators: (i) the mining industry's contribution to local development indicator (MILDI). This indicator is defined as the share of the turnover of the mining companies destined to local development and local communities' well-being improvement. The percentage of this local tax is generally defined in the mining regulations Acts of each mining country. (ii) As argued above, a household Purchasing Power Indicator (PPI) is a good indicator of the impact of mining operations on local household's life conditions on the income increase point of view. (iii) The Rural Development Indicator (RDI): Literature shows that mining

---

[5]Reader can find more details on mining case studies in Carney (1998), Hausmann and Rigobon (2003), Hugon (2009), Hilson (2010), Andersen and Aslaksen (2008), Van der Ploeg (2011), Arezki et al. (2012), Jensen and Wantchekon (2004), Auty (2001), Chen et al. (2010), Papyrakis and Gerlagh (2004), Bolt et al. (2005), Sba-Ecosys-Cedres (2011), Reed et al. (2006), Kaufmann et al. (2009), Brollo et al. (2010), Frankel (2010), Oxfam (2013), Geiregat and Yang (2013), Bardhan (1997), Dialga (2015).

[6]Lu et al. (2017) argue that methods for assessing interactions between different parameters of Sustainable Development (environmental, demographic, social and developmental aspects) were not sufficiently developed or applied. In order to improve this situation, Local Sustainable Development Indicators need to be developed to provide a solid base for policy-making and to measure progress toward political objectives. Therefore, to achieve global sustainability, acting at the local level is important.

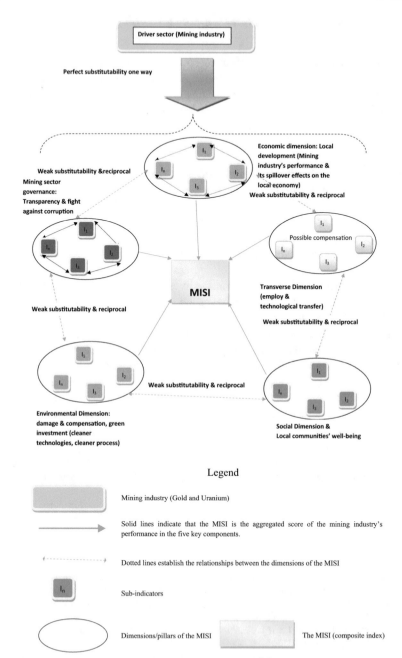

Fig. 1 MISI's theoretical framework

revenues can serve to stimulate rural sector (agriculture and livestock) as this sector has been identified as a sector that could be more likely to employ a larger share of the labor force and is able to replace the mining sector after the depletion of the resource. Authors like Irz et al. (2001), Cadot et al. (2013a, b), Gamu et al. (2015) pointed out the spillover effects of that sector. The RDI is defined as the share of the government mining revenues devoted to the rural development fund and in proportion to the rural population regarding the total population of the country.

3.2.2 Social Dimension and Local Communities' Well-Being

As it is known that the extraction of mines, Uranium as well as Gold will take end, it is crucial to define a clear rule to guarantee future generations their abilities to live in mining countries. To this purpose, we define an Intergenerational Equity Indicator (IGEI). Furthermore, it showed that mining industry rises inequalities across regions and across communities (those dispose rights to receive mining rents are more likely to capture an import part of the rent than the rest of the population). In order to take into account this negative effect, we define an Intra-generational Equity indicator (IEI). Finally, Children well-Being's Indicator (CBI) is defined to take into account children labor and their life conditions, mainly in artisanal mining sites. The IEI uses the CPIA—Country Policy and Institutional Assessment indicators from the World Bank. They reflect the efforts in equal opportunity (access to the labor market, justice, public services) and gender equality. The IGEI is defined as the ratio between expenditure and investments (from mining revenues) for immediate needs and long-term investments.

3.2.3 Environmental Dimension: Damage and Compensation, Green Investment (Clean Technologies, Cleaner Process)

The mining industry is very often decried because the sector would induce significant environmental damage, some of which may even exceed irreversibility thresholds. In the context of corporate social responsibility, many of the mining companies are investing in either offsets, cleaner technologies, or even the process of extracting the mineral resource, in order to reduce the impact of mining activity on the environment. It is important to make tangible such efforts within the framework of the sustainable mine. We define two key environmental indicators. The first one is a Net Environmental Impact Indicator (NEI) and the last one is an Industrial Ecology Indicator (IEcoI). More and more, mining industry adopts the industrial ecology as a part of its business model. The industrial and territorial ecology is a method of production rooted in the territory which consists of seeking eco-industrial synergies at the scale of a business area; the waste of one company that can become the resources of another. The NEI is the ratio of green[7] investment to environmental damage caused by mining. Bai et al. (2017) argue that investing in green technologies has important positive impacts on environmental footprints and for environmentally sustainable practices.

Marin et al. (2016) demonstrated how mineral resources and reserves can be managed to achieve a sustainable form of small-scale mining (SSM), based on the concepts of proving a "minimum reserve" and working with "replication" of the operation on subsequent small reserves.[8]

The Industrial Ecology Indicator is defined as the share of recyclable waste (materials) used as input in the mining operations[9] following the concept of industrial ecology.

3.2.4   Transverse Dimension: Employment and Technological Transfer

The notion of "transverse" dimension refers to the effects of mining activity on several dimensions of development at the same time. In other words, some actions of the company may have other effects beyond the sole objectives. We talk about overflowing effects or positive externalities. That is typically the case of techno-logical transfer through the Foreign Direct Investment (FDI) (most of large-scale industry's investment comes from the FDI). Employment has also both economic and social impacts. Indeed, Employment is at the crossroads of the economic and social dimensions (Kotsadam and Tolonen 2016). Employment is a mean of self-fulfillment (Dubois 2009; Ballet et al. 2012; Clément et al. 2012). Employment also has an economic character when it is perceived as a workforce, and therefore an input for the company.

Given all these reasons, we identified two indicators to take into account the overflowing effects. The first indicator is the mining sector's Local Employability Indicator (LEI). This indicator is defined as the share of local communities in the workforce of employees of mining companies. The second indicator is the Technology Transfer Indicator (TTI). Mining industries are an excellent vector of technology transfer, the effects of which benefit the entire national and local

---

[7]Green investment refers to investment in cleanup technologies, resource optimization used in the mining extraction, wastewater treatment, alternative technology to cyanide and mercury, etc. For example, green ITS has the capacity not only to help minimizing energy consumption of orga-nizations and mines (Kusi-Sarpong et al. 2016), but also to support in mitigating the overall environmental impact significantly (Ryoo and Koo 2013).

[8]The *minimum reserve* is the volume of mineral whose exploitation allows for the payment of the initial investment, the operating costs, the cost of mineral exploration needed to extend the proven reserve, plus the desired profit. *Replication* is the exploitation in cycles of several volumes of minimum reserves (Marin et al. 2016).

[9]See Caiado et al. (2017), Heravi et al. (2015) through their eco-efficiency analysis or Gandhi et al. (2018) through their lean and green manufacturing concept.

economy provided that connection canals are well thought out (establishment, upstream and downstream, of industries using similar technologies, training and technology transfer plan, skills upgrading for local employees, etc.). In addition, the access to technology enables more environmentally friendly production (environmental dimension). This point is highlighted by Ryoo and Koo (2013) and most recently confirmed by Bai et al. (2017). The economic effects of a technological transfer are also evident: time savings, optimization of the use of resources, improvement of working conditions, etc. Access to technology also improves governance by e-payments of mining taxes. Citizen could also check the use of mining revenues by the governments (transparency). As the technology transfer is driven by the FDI and as the mining industry is the main contributor of the FDI flows in these countries, we define the TTI as a proxy of the share of the FDI in the country's GDP.

3.2.5 Mining Sector Governance: Transparency and Fight Against Corruption

The mode of governance of the resource is crucial in a local sustainable development process based on mining. The conditions of profitability of the mining rent, redistributions and thus equity will depend on the degree of transparency in the management of the resources in exploitation. It seems obvious that opaque management by clans will have negative effects on the other development factors highlighted.

The *Natural Resource Governance Institute*[10] (NRGI) has developed an indicator to assess five governance domains related to natural resources in exporting countries from 80 to 90% of the world's natural resources. The five levels of governance considered by the NRGI are the country's institutional and legal framework, resource disclosure practices, guarantee and quality control measures, conditions governance, and transfer levels from resource management to local communities.

At the same time, NGOs such as *Global Witness* and *Extractive Industries Transparency Initiative, (EITI)* are working with mining companies and mining country governments to make the mining sector transparent. The work of the EITI is to verify the compliance of mining tax payment declarations. The EITI confronts the mining companies' statements separately with the amount of mining taxes that governments claim to have received from the mining companies. The NGO then makes a consolidation work to highlight any gaps. This component of the index is subdivided into three indicators: A Mining Governance Indicator (MGI) is defined as assessing the mining industry management method in the two countries (Burkina Faso and Niger). Data for this indicator come from the Natural Resource Governance Institute publications. A Mining sector Transparency Indicator

---

[10]http://www.resourcegovernanceindex.org, last access: 16/07/2018.

(MTI) is built from the consolidation's work of the EITI. Scores are attributed to each mining sector (Gold sector and Uranium sector) proportionally to the gap of declarations of mining revenues payments revealed by the EITI. The Control of Corruption Indicator (CCI) is defined to measure the willingness and the ability of the governments to fight against corruption in the mining sector in particular and more generally in the whole society. The synthesis of the MISI's dimensions and the associated individual indicators is given in Table 7 in the Appendix.

4 Data Collection, Sources and Periods of Study

To implement the MISI, we used both primary and secondary data.[11] Scores are calculated for three periods: 2010, just after the expansion of mining industry in Burkina Faso (2015) as the reasonable date to get feedback and to assess the real impact of mining industry activities on local development and communities. 2017 is an update of 2015 index scores. Primary data are deduced from our field survey. Weights are directly generated by our respondents via the Budget Allocation Process described later. Indeed, one of the field survey's aim was to assign weights to the selected indicators by the various stakeholders concerned by the mining operations. The score corresponds to the relative importance of the issue in terms of local sustainable development in connection with mining operations according to the expectations of local communities. Our case study is conducted in Burkina Faso (for the Gold sector) and Niger (for the Uranium sector). To ensure the representativeness of our sample, we matched it with the national statistics: in Burkina Faso, we had 53% of women, 46.2% of young people, 86% of people living in rural areas.[12] In Niger, the proportion of women is 50.6%. Young people account for 49.2% of the total population and 79% of Nigeriens lived in rural areas and have land work as the main occupation.

We also used secondary data got from both national and international statistics institutes as well as the reports of mining companies.[13] The main sources utilized are The Burkina Faso Statistics and Demographic National Institute, the Niger National Statistics Institute, the Statistics Department of Mine, Petroleum Ministry and the Ministry of Economics and Finance of both countries. Data from the World Bank, the Revenue Watch Institute, the Statistical Yearbook, Extractive Industries

---

[11]Primary data: data from field survey; Secondary data: data compiled by national or international institutions or statistical data from government administrations and mining companies' reports.

[12]Data sources: National Institute of Statistics and Demography (NISD-Burkina), NIS-Niger.

[13]Kiaka Gold Sarl (2014); SOMAIR-Société des Mines de l'Aïr (2012); SOMINA-Société des Mines d'Azélik (2014); COMINAK-Compagnie Minière d'Akouta (2012); Plan African Minerals LTD (2014); SEMAFO Burkina SA (2013); Midas Gold Sarl (2013); Endeavour mining-Avion Gold Burkina Sarl (2014); Bissa Gold SA (2014); Ampella Mining Limited (2013); IAM Gold Essakan SA (2013); Gryphon Minerals Limited (2014).

Transparency Initiative, Transparency International, Economist and Intelligence Unit are also included in the study.

Some environmental data are estimated through the formula described below. The estimated data are based on the yearly reports of mining companies and some previous studies like those of Leduc and Raymond (2000); Field et al. (2000); Levine et al. (2007); Sba-Ecosys-Cedres (2011); Adiansyah et al. (2015); Caiado et al. (2017); Heravi et al. (2015); Gandhi et al. (2018). Most of these studies dealt with the challenge of the environmental cumulative damages assessment. Indeed, we agree with Adiansyah et al. (2015) that the most critical issue in mining operations and more specifically in tailings management is the taking into account the irreversible impacts of tailings. The equation below takes into account the cumulative effects of environmental damage attributable to the mining industry.

$$I_t = \sum_{t=0}^{n} \frac{a_t \Delta I_t}{(1+r)^{-t}} = \sum_{t=0}^{n} a_t \Delta I_t (1+r)^t \tag{1}$$

where:

I_t is the cumulative impact at t; ΔI_t the impact attributable to the operations of period t;

r is a capitalization rate.[14] We choose to index the capitalization rate to public investments rate because environmental assets are considered as public goods and that way, the environmental impacts are assessed in monetary terms (current value of the currency). So it seems important to capitalize the amounts of past impacts by using public investments rate.

$$a_t = 1 - \frac{effort\ to\ reduce\ the\ impact\ of\ year\ t}{generated\ impact\ at\ t} = 1 - \frac{reduction\ of\ \Delta I_t}{\Delta I_t} \tag{2}$$

a_t is the net impact accumulation, i.e. the cumulative impacts of past years on the current environment. Two interpretations of the coefficient a_t are possible: when at is 1, that means that the environmental damage of the previous years are fully added to those of the current year. In contrast, a_t is equal to zero means that companies make efforts to invest in damage compensations.

---

[14]When it is to assess the future costs, r is defined as the discount rate.

5   Normalization, Weighting, and Aggregation

5.1   Normalization

As initial variables have different units of measurement, it is necessary to standardize these units of measurement because one cannot add pears to apples! However, as we defined performance ratios for the MISI's construction (see Sect. 2.2), this normalization step is no longer necessary. Our sub-indicators are defined as follows:

$$SI = \frac{V_{ij}}{V_{\text{max}j}} \tag{3}$$

SI is the normalized sub-indicator, V_{ij} is the value of the initial variable for the entity i and $V_{\text{max}j}$ is the maximum value reached by j. The maximum value can also be chosen from the sample, by using the best in the class criterion. Values of standardized indicators range from 0 to 100. Such a scheme of scale is easy to interpret the score because 100-normalized score indicates the percentage of effort which remains to reach the maximum performance, or the performance of the best in the class.

5.2   Weighting

After normalizing the sub-indicators, the next challenge is how to assign weights to each indicator and how to justify the choice of one method than another. Assigning different weights to sub-indicators is justified in the literature (OECD and JRC 2008; Dialga and Le 2017) as issues can be differently appreciated by the researcher. In addition, the importance of the issues can also depend on the expectations of stakeholders. However, each choice of weighting should be justified. Indeed, how to legitimize weights associated to the indicators in a scientific approach? For example, choosing equal weighting system has the merits to be simple and so transparent. However, this method is the most criticized because assigning equal weights to all indicators is an arbitrary choice. Another weighting system which is more and more used in literature is the Principal Components Analysis (CPA) weighting system. The PCA method enables generating weights based on statistical computations. However, PCA method requires a sufficient number of observations, usually time series. Then the data of our study (cross-sectional data) are not sufficient for such a method.

The weighting method employs in this study is the so-called Budget Allocation Process (BAP). As already explained above, the BAP method enables the gathering of opinions from various stakeholders. In our case study, stakeholders include experts in local development, NGOs, local governments, local communities and mining companies. Our choice is mainly guided by our intention to offer both scientific and professional legitimacy to the MISI. The approach has the merit of guaranteeing the local acceptability of the decision support tool[15]. However, the method does not escape classical criticisms that are addressed to composite indices. Indeed, Dialga (2018) has been faced with the following criticisms: How can the viewpoints of stakeholders with competing interests converge? How can stakeholders be engaged in order to reach a consensual compromise? How can we be sure that the points of view of stakeholders are coherent? How can the results from such a participatory approach be objectified? In order to bring some answers to these questions, Dialga (2018) calculated a consistency index[16] of appreciation to objectify the BAP method. We replicate this calculation of this consistency index in this chapter as we use a part of Dialga (2018)'s data.

Table 2 summarizes the results of the Budget allocation Process. Values are multiplied by 100 in order to facilitate the interpretations. Globally we can subdivide the themes submitted to the stakeholders into five groups: the issue of skills' transfers and health's improvement is largely preferred by the respondents than the others issues. The respondents devoted 13% of their budget to this theme. The second group of issues includes transparency and fight against corruption, intergenerational equity and children life's conditions and employment and technological transfer. Respondents agreed to allocate 12% of their budget to each of these issues, one percentage point less than the issue of skills' transfer and the health of local communities. According to stakeholders, these four issues have the same importance and should be equally weighted in the MISI. The third group includes the issue of local development and local communities' well-being. This issue should be taken seriously, immediately after those of the second group identified by the respondents because stakeholders allocated 11% of the budget to this issue. The fourth group of issues in connection with sustainable mining industry and local development includes the mining industry's performance, its spillover effects on the local economy, its impacts on environment and the issue of sustainable management of the mines. These issues account for 10% in the stakeholders' budget allocation.

The last group includes the issue of intra-generational equity. This theme got the lowest weight for several reasons. This weighting reflects the distribution of mining

---

[15]See Bayulken and Huisingh (2015); Hodge (2014) and Lu et al. (2017) for discussions about local legitimacy of SD indicators. For example, even if the top-down approach often contains quantitative indicators as a way to evaluate the broader context of sustainable development (Turcu 2013), these indicators tend to ignore local issues from local perspectives. Experiences learnt from van Zeijl-Rozema and Martens (2010) seem to show that EU indicators failed to measure sustainability at the regional level because of the lack of local communities' representativeness.

[16]$I = x \frac{w_j}{w_{j'}}$; With x the ratio of budget attributed to sub-indicators s_{ij} and $s_{ij'}$; w_j and $w_{j'}$ are the weights of indicators s_{ij} and $s_{ij'}$, obtained from the Budget allocation. Decision criterion: $I \leq 10\% \Rightarrow$ consistent judgment.

Table 2 Summary of BAP weighting method

	Skills' transfer and health's improvement	Transparency and fight against corruption	Mining industry's performance	Intra-generational Equity	Intergenerational Equity	Environment and sustainable management of mines	Local development and local communities' well-being	Children life's conditions	Employment and technological transfer
Obs.	164	164	164	164	164	164	164	164	164
Min	2.6	0.0	0.0	0.0	2.0	0.0	0.0	2.2	2.2
Mean	**13**	**12**	**10**	**9**	**12**	**10**	**11**	**12**	**12**
Max	23.7	20.9	20.9	25.0	22.2	21.0	17.9	23.1	20.5
St. Deviation	22.7	3.5	3.6	3.8	4.0	2.7	4.0	3.3	3.0

Consistency index: 0.0

Note final weights are calculated from the BAP results. However, we grouped the themes above following the components of the MISI and then, the weights associated with the dimensions are calculated in proportion to the themes that make up each dimension of the index (see the final weights in appendix) income that benefits to a small number of individuals. It also reflects the social inequalities observed in the society that the respondents wanted to highlight.

Overall, we note that the dimensions of the MISI do not have the same importance according to the respondents. This result legitimizes our choice to abandon the equal weighting system although this method is more transparent. From the point of view of scientific legitimacy, the choice of the BAP method is validated by the consistency index (see Table 2) that we have calculated.

5.3 Aggregation

The aggregation aims to condense the information of initial variables in order to obtain a single number. There are several methods of aggregation and each method can lead to different results. So the crucial question for the researcher therefore is: what method should be chosen and how to justify the chosen aggregation method (Dialga and Le 2017)? Let's summarize here three methods most commonly used. We take care to highlight both the strengths and the limits of each of them.

5.3.1 Linear Aggregation

The linear aggregation method is defined by the following equation below:

$$CI_i = \sum_{j=1}^{n} w_j I_{ij}$$

$i = 1, \ldots m$ entites; $j = 1, \ldots, n$ variables; CI, composite score;

(4)

I is the initial normalized variable and w is the associated weight.

As the equal weighting method, the linear aggregation method is simple and transparent.

Composite indices are sometimes criticized, hated and rejected because they are considered as "black boxes". Choosing the linear aggregation method has the merit to avoid or to reduce the criticism related to the complexity of composite indexes' building.

However, choosing such a method is to take the risk of violating the perfect substitutability hypothesis (Zhou et al. 2006) which is nevertheless constantly controversial (Munda and Nardo 2003, 2009; Munda 2005). The method also raises problems in interpreting the weights associated with the sub-indicators. According to Munda and Nardo (2003), weight associated to the sub-indicators in the linear aggregation method should be interpreted as "compromise coefficients" among dimensions or component of the composite index. However, the compromise coefficients are different from the coefficients of importance that the weights would reflect. The choice of a linear aggregation method is more problematic in the issue of a sustainable mining industry as we have conflicting dimensions. How and to what levels can we admit the depletion of natural resources and pollution offset by gains expected for the local development and local communities' well-being improvement? In this regard, we can remember the remark of Munda and Nardo (2009): "*Complete compensability implies that an excellent performance on the economic dimension can justify any type of very bad performance on the other dimensions, which is exactly what the concept of sustainability tries to avoid*" (Munda 2005).

5.3.2 Non-compensatory Aggregating Method Under Perfect Complementarity Hypothesis

This method is an application of the multi-criteria approach to the composite indexes field. The method requires at least two individuals (from statistical point of view) to generate scores. Let j and k two individuals. The composite score of j, given the performance of entity k and considering M individual indicators is given by the following equation:

$$e_{jk} = \sum_{m=1}^{M} \left[w_m \left(P_{jk} \right) + \frac{1}{2} w_m \left(I_{jk} \right) \right] \tag{5}$$

$w_q \left(P_{jk} \right)$ and $w_q \left(I_{jk} \right)$ are individual indicators' weights. P_{jk} and I_{jk} denote a preference and indifference relation, respectively, for j and k. The final score of multiple possible rankings is obtained from the sum of the pairwise comparison scores.[17]

$$\phi^* = \max \sum e_{jk} \tag{6}$$

The method is appropriate for dealing with issues where the researcher is required to seek compromise solutions. This situation arises when stakeholders have irreconcilable positions. In other words, these are situations where it is impossible to reach consensus on all the issues. The method also admits the joint use of both qualitative and quantitative data (OECD and JRC 2008, p. 115). The method is also consistent with ratios of performance or interval scales normalization method. Bouyssou (1986) and Bouyssou and Vansnick (1986) emphasize the local legitimacy of the method because the process involves stakeholders, usually chosen by local communities.

---

[17]For a full and detail description of this method see Munda and Nardo (2009).

However, there are also limits to this method. From the theoretical point of view, this method is based on a strong hypothesis, namely the hypothesis of perfect complementarity. This hypothesis implies that any individual who wishes to improve his composite score should perform at the same time on all dimensions of the index; which is hardly feasible in a real world. Furthermore, as mentioned above, the method requires at least two individuals for the score calculation. But the method also becomes technically ineffective when the number of individuals becomes high.

5.3.3 Geometric Aggregating Method

The geometric aggregating method is defined as follows:

$$CI_i = \prod_{i=1}^{n} I_{ij}^{w_j} \tag{7}$$

$$i = 1, \ldots, m \, entities; j = 1, \ldots n \, variables$$

The geometric aggregating method admits imperfect substitutability among the components of this index. This is the main idea highlighted by the theoretical framework (see Fig. 1) given the conflicting issues raised by the mining sector. From the robustness point of view, Zhou et al. (2006) support that the geometric method is preferable to other aggregating methods because the geometric one is less sensitive to extreme values.

However, and more generally, the method pulls the score of individuals down. From this point of view, the geometric aggregation does not sufficiently highlight the efforts made by the mining companies. In addition, before choosing the geometric method, we must ensure that the available data satisfy the statistical properties required by the geometric aggregation (no null values and no negative values in the data).

Learning from the strengths and weaknesses of each method and referring to the theoretical framework of the MISI, we retain the geometric aggregation method. However, the linear aggregating method presented in this chapter will be used to assess the robustness of the index.

Finally, the MISI is calculated via the following function:

$$MISI_i = \prod_{i=1}^{5} (SI_i)^{w_i} \tag{8}$$

SI_i is the normalized sub-indicator, w_i is the associated weight.

Table 3 Gold and Uranium sectors sustainability index's scores (2010, 2015, 2017)

	2010 MISI	2015 MISI	2017 MISI
Gold sector	38.80	39.93	48.45
Uranium sector	38.96	41.38	42.07

Table 3 shows the mining industry's performances in terms of sustainability for three periods: 2010, 2015, and 2017. According to our criterion of sustainability, defined above, neither of the two mining sectors is sustainable in these two countries because none of them reached the score of 67 out of 100. However, both sectors make year-to-year efforts and converge toward a sustainable path. Indeed, the score of the gold sector in Burkina Faso has increased from 38.80 out of 100 in 2010 to 39.93 in 2015 and 48.45 out of 100 in 2017. At the same time, the uranium sector in Niger has followed the same trend. Indeed, the MISI score was 38.96 on a scale of 100 in 2010, 41.38 in 2015, and 42.07 in 2017.

However, as highlighted in Fig. 2, convergence toward the sustainable path is faster in the gold sector than in the uranium sector. While the MISI score for the gold sector was lower than the MISI score for the Uranium sector at the beginning of the period (38.80 versus 38.96 in 2010), the score in the sector of gold is higher than that of uranium at the end of the period (48.45 versus 41.07 in 2017). These results suggest that it is less difficult to comply with the requirements of corporate social responsibility in the gold sector than it is for the uranium sector.

In order to better understand the mining industry sustainability performance, we make a cross-section analysis. We focus on the performances of the more recent year, i.e., performances in 2017.

Figure 3 shows that, apart from the cleaner production method, the two mining sectors have roughly the same performance profiles in terms of the sustainability of the mining sector. The two mining sectors have a strong impact on the well-being of local communities according to the results highlighted in Fig. 3 (a score of 59 and 57 for gold and uranium sectors, respectively).

However, the impact of mining is limited on local development (a score of 36 and 32 on a scale of 100 induced by the gold sector and the uranium sector, respectively), despite the good structuring effects highlighted in both sectors. Indeed, the score of the structuring effects is 48 out of 100 in the gold sector and 51 out of 100 in the uranium sector. Transparency and good management practices are significantly higher in the gold sector than they are in the uranium sector (see Fig. 3). Finally, it seems that the difference in the cleaner production method between the two sectors can be explained by the fact that tailings of the uranium sector are more expensive to be treated through the recycling technique (ecological industrial approach) and present enormous risks because of the effects related to radioactivity.

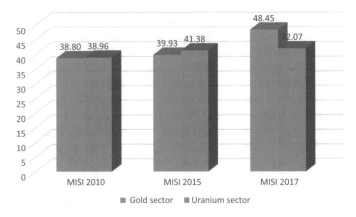

Fig. 2 Sustainability performances in Gold and Uranium sectors

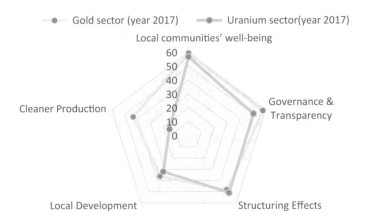

Fig. 3 Mining industry sustainability performance by component

6 Sensitivity and Robustness Analysis

Sensitivity and robustness analysis aims to assess the strength of the index. To do so, we calculate the variations of the scores of the index following change in both weighting and aggregation methods. For this purpose, we consider three methods of weighting (equal weighting, weighting by PCA and weighting by Budget Allocation Process (our baseline)). We combine these weighting methods with the geometric aggregation method and linear aggregation method. Due to the particularity of the aggregation method by the multi-criteria approach highlighted in the previous section, we exclude this method from the robustness analysis of the index. Comparing the scores generated by the multi-criteria method with other aggregation methods could lead to misinterpretations.

We then test the robustness of the index. We use the measure of Shannon-Spearman lost information proposed by Zhou et al. (2006); Zhou and Ang (2008). The Loss Information Measure (LIM) aims to evaluate the ability of the composite index to return the information contained in the individual variables. The third step aims to analyze possible links that may exist between the new index and some other indexes. We conduct a correlation test between the MISI and the SIMC of Dialga (2018), as well as between the MISI and the Resource Governance Index (RGI) of the Natural Resource Governance Institute (NRGI). Checking the existence of possible links between these three indices seems to us relevant for the reason that all the three indices deal with the same issue, with the only difference being the different scales of analysis (sectoral analysis for the MISI and country-level analysis for the SIMC and the RGI).

6.1   Sensitivity Analysis

6.1.1   Sensitivity Analysis by Variance Decomposition

As described by Homma and Saltelli (1996); Cherchye et al. (2008); Saisana and Saltelli (2010) and adopted by Aguna and Kovacevic (2010); Dialga and Le (2017) and Dialga (2018), the sensitivity analysis by variance decomposition is used to assess the contribution of each input variable (dimensional index) in the formation of the total output (composite score). The contribution of each individual indicator is given by the following equation:

$$S_j = \frac{V_{X_j}(E_{X_{-j}}(Y \backslash X_j))}{V(Y)} = \frac{Vj}{V(Y)} \tag{9}$$

S_j gives the relative contribution of the jth variable (only) to the total variance. Thus, the more important the area or the dimension defined by variable X_j is, the greater S_j will be. In particular, when the variable explains almost all variations of the output, the sensitivity indicator tends toward unity.

The variance decomposition of the MISI score is highlighted by Fig. 3. Results of Fig. 3 show that all the five components of the index are relevant as they contribute in different proportions to explain the variations of the index score. The component named "Governance and transparency" is those more explains the variations of the MISI score (30.74% of the index score variation). This component is followed by those named "structuring effects of the mining industry" (29.54% of total variance). "Cleaner production techniques" and "impacts of mining industry on local development" have almost the same contribution to the index score variation (24.05% and 23.79%, respectively). Finally, the component "impact of mining industry on the local communities' well-being" has an intermediate

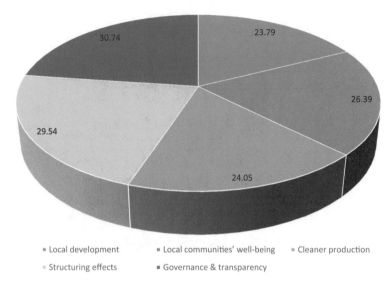

Local development
Local communities' well-being
Cleaner production
Structuring effects
Governance & transparency

Fig. 4 Variance explained by the components of the MISI

contribution (26.39% of index score variation). Results of variance decomposition suggest that, in order to improve the MISI's score, efforts should be directed first and foremost to issues of governance and transparency in the extractive industry and those relating to the structuring effects of the mining industry and then to the other dimensions of the index afterward.

6.1.2 Sensitivity Analysis by Changing the Aggregating and Weighting Methods

Figure 4 shows that the MISI score is not stable and changes according to the aggregation-weighting method used, regardless of the mining sector. The Linear-Budget Allocation Process (LIN-BAP) and Linear-Equal (LIN-EQUAL) aggregation-weighting systems are favorable to the score of the index while the Linear-Principal Components Analysis (LIN-PCA) and the Geometric-Principal Components Analysis (GEO-PCA) aggregation-weighting systems derive the index score downwards compared to our baseline (Geometric-Budget Allocation Process method).

This instability of the scores, following the changes of methods, is not a surprising result and is even a specific characteristic of the composite indices. Previous studies highlighted the unstable nature of several composite indices (see Desai 1991; Coste et al. 2005; Zhou et al. 2006; Brand et al. 2007; OECD and JRC 2008; Cherchye et al. 2006; Nathan et al. 2012; Chinn and Ito 2008; Tarantola and Vertesy 2012; Perišić 2015; Dialga and Le 2017). We are aware that the MISI will not escape this traditional criticism. However, there is no reason to be alarmed

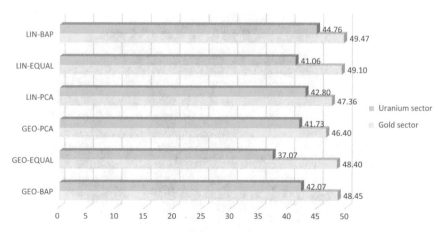

Fig. 5 Changes in the 2017 MISI's score following the weighting and aggregation methods

Table 4 MISI'score variation matrix

	GEO-BAP	GEO-EQUAL	GEO-PCA	LIN-PCA	LIN-EQUAL	LIN-BAP
GEO-BAP	0.00					
GEO-EQUAL	0.15	0.00				
GEO-PCA	2.05	1.90	0.00			
LIN-PCA	1.09	0.94	−0.96	0.00		
LIN-EQUAL	−0.65	−0.80	−2.70	−1.74	0.00	
LIN-BAP	−1.02	−1.17	−3.06	−2.10	−0.36	0.00

because the variations of the MISI's score are in reasonable proportions and can be defined over a confidence interval of 0.15 to 3.06% in absolute value. Indeed, using the 2017 MISI's score, we calculated the gaps of scores according to the selected weighting and aggregation methods. The results of the calculation are reported in Table 4. It can be seen that the variations of the score are small. In conclusion, the assessment of the robustness of the MISI must be based on the associated confidence interval.

6.2 Robustness Analysis: The Loss Information Measure (LIM)

The coefficient of adequacy of the aggregation method is a measurement of the quantity of information reveled by the composite score compared to those contained in the original variables. The Loss Information Measure (LIM) as a measure of the aggregation method's adequacy, in the field of composite indicators, is proposed by

Zhou et al. (2006). The LIM aims to assess the appropriateness of a method of aggregation of the CI over another. The LIM is defined as follows:

$$LIM = \left| \sum_{j=1}^{n} w_j(1 - e_j)r_{sj} - (1 - e)r_s \right| \tag{10}$$

w_j is the associated weight,

$$p_{ij} = \frac{X_{ij}}{\sum_{i=1}^{m} X_{ij}} \tag{11}$$

$i = 1, \ldots, m$ entities (e.g : companies); $j = 1, \ldots, n$ variables

$$p_k = \frac{I_k}{\sum_{i=1}^{m} I_k} \tag{12}$$

$$k = 1, \ldots, m$$

p_{ij} and p_k are, respectively, the normalized initial variables and the composite score.

$$e_j = -\frac{1}{\ln m} \sum_{i=1}^{m} p_{ij} \ln p_{ij} \tag{13}$$

$$j = 1, \ldots, n$$

$$e = -\frac{1}{\ln m} \sum_{i=1}^{m} p_k \ln p_k \tag{14}$$

e_j and e are performance measures for each entity. According to Zhou et al. (2006), Zeleny (1982) shows that $0 \leq e_j, e \leq 1$.

Following the LIM decision criterion, Table 5 shows that the best choice is the geometric aggregation and weighting by the Budget Allocation Process. Indeed, by choosing this combination, we only lose 7.7 and 13.5% of the information contained in the initial variables for the uranium sector and gold sector data, respectively. These information losses are lower compared to other combination choices. For example, choosing the combination "linear-equal" would lose 18 and 12% of information, respectively, for the gold and uranium sectors.

So, choosing the geometric aggregation method and BAP minimizes the loss of information (the loss that can be avoided) by 56% on uranium sector data and 33% on gold's industry data compared to the example above.

Table 5 Loss Information Measure values according to the selected weighting and aggregating technique

Aggregation	Weighting					
	Equal		BAP		PCA	
	Gold	Uranium	Gold	Uranium	Gold	Uranium
Linear	18.0	11.8	15.1	8.3	17.1	12.1
Geometric	15.6	10.6	**13.5**	**7.7**	15.2	9.8

Note The LIM's values are replicated from Dialga (2018) who used similar data and the same tool to assess the robustness of his index. We dropped results of the multi-criteria method because of its particularity compared to our selected methods. In Table 5, we report the values of the Loss Information Measure (LIM). As defined by Dialga (2018), *the LIM is the difference between information contained in initial variables or sub-indicators and those given by the composite score, given a selected combination of methods of aggregation and weighting. The best aggregation-weighting method to be retained is that linked to the lowest value of LIM.* This best method is the "Geometric-BAP" aggregation-weighting method (see the associated LIM-values in bold)

Table 6 Correlation matrix between MISI, SIMC, and RGI

	MISI	RGI	SIMC
MISI	1.0000		
RGI	0.968	1.0000	
SIMC	0.944	0.987	1.0000

6.3 Link with Others Sustainable Development Indices Analysis

This sub-section aims to establish possible links between the MISI and similar indices through a correlation analysis. We selected the following composite indices because of their similarities: the MISI, the Sustainability Index of Mining Countries (SIMC) of Dialga (2018) and the Resource Governance Index (RGI) of the Natural Resource Governance Institute (NRGI).

The correlation test summarized in Table 6 shows that the three indices related to the mining industry sustainability are highly correlated. This is not a surprising result in view of the issues taken into account in each index.

As the correlation coefficients are all positive, this means that the selection of the indicators for the MISI's construction was judicious because these indicators would evolve in the same direction as the variables selected of the RGI and the SIMC.

This result confirms the relevance of the MISI and legitimizes its ability to describe sustainability in the mining sector in the same way as the recognized indices in this field, notably that of the Natural Resource Governance Institute and the Dialga (2018) index.

However, there is a difference in score for the same mining sector, both the gold sector and the uranium sector, depending on the index considered. Indeed, although

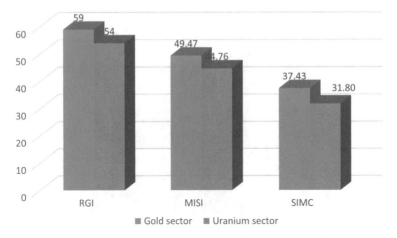

Fig. 6 Mining industry sustainability performance in 2017 according to three selected indexes

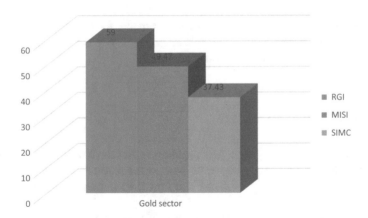

Fig. 7 Sustainability performance in Gold sector in 2017 according to three selected indexes

we followed the same construction techniques of the three indices (linear aggregation, unequal weighting), Figs. 6, 7, 8 show significant differences for the same mining sector in terms of sustainability by considering the scores of the three indices (MISI, SIMC, and RGI). There are three main reasons: First, even if the construction techniques of the indices are identical, the values assigned to the initial variables differ according to the sources of data collection. Indeed, the Natural Resource Governance Institute, which collects data through surveys, tends to give high values to variables compared to values reported on the World Bank or Transparency International websites for the same variables. Second, the greater or lesser number of initial variables considered in the construction of the composite index can influence the composite score of each index. The three selected indices do

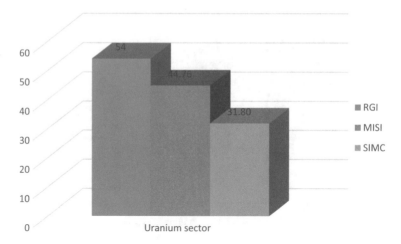

Fig. 8 Sustainability performance in Uranium sector in 2017 according to three selected indexes

not have the same number of initial variables. Finally, the weights associated with each component of the composite index are not identical in the three indices. These unequal weights induce different composite scores.

7   Conclusion

This chapter aimed to develop a Mining Industry Sustainability Index in order to show how does the mining sector contribute to the expansion of other economic sectors, local communities' well-being and how does it manage its environmental impacts?

Based on our sustainable mining industry rule, i.e., on a scale of 100, or considering the core components of this index, the mining sector which obtains more than 67 of composite score or which performs more than 50% in 2/3 of the five components of the index is considered as sustainable, neither the gold sector nor the uranium sector is sustainable in 2017. These mining sectors were not sustainable either in 2015 or in 2010. However, the two sectors are converging on a sustainable path at an appreciable pace when we refer to their general trend.

The sensitivity and robustness analysis, using the variance decomposition method and the loss information measure of Zhou et al. (2006) and the correlation tests with other well-known indicators, at the end of the chapter, proved the strength of the sustainability index of mining industry. However, the index will not escape criticism linked to the instability of composite scores following changes in methods. That is why we defined a confidence interval associated with index scores. The assessment of the robustness of the MISI should be based on the associated confidence interval.

The results should be considered with caution as the study has some other limitations including limits related to the low quality of data and the uncertainties in the estimation techniques for missing data. A joint multidisciplinary research project, including disciplines as Law, Sociology, Economics, would be necessary for a complete study on mining sector in that countries. Such a research project should provide a substantial budget for the collection and production of reliable statistical data.

Appendix

Table 7 Summary of the MISI construction: components, sub-indicators, weights and data sources

Component	Indicators	Indicator abbreviation	Weighting methods			Data sources
			BAP	PCA	Equal	
Local development: Mining industry's performance and its spillover effects on the local economy	Mining industry's contribution to Local development Indicator	LDI				NRGI
	Household Purchasing Power Indicator	PPI				WDI
	Rural Development Indicator	RDI				Ministry of Finance
	Local Development Index	**LDI**	**28.57**	**42.7**	**20**	
Social Dimension and local communities' well-being	Intra-generational Equity indicator	IEI				WDI
	Intergenerational Equity Indicator	IGEI				WDI
	Children well-Being Indicator	CBI				WDI
	Local communities' Well-being Index	**CWI**	**36.51**	**2.8**	**20**	
Environmental Dimension: damage and compensation, green	Net Environmental Impact Indicator	NEI				BUNEE-BF; BEEI-NIGER
	Industrial Ecology Indicator	IEcoI				

(continued)

Table 7 (continued)

Component	Indicators	Indicator abbreviation	Weighting methods			Data sources
			BAP	PCA	Equal	
investment (cleaner technologies, cleaner process)						Mining companies' annual reports
	Cleaner Production Index	**CPI**	**7.94**	**0.5**	**20**	
Transverse Dimension (employment and technological transfer)	Mining sector Local Employability Indicator	LEI				CMB; DGMG-BF; DGMG-NIGER
	Technology Transfer Indicator	TTI				WDI
	Structuring Effects Index	**SEI**	**17.46**	**14.6**	**20**	
Mining sector governance: Transparency and fight against corruption	Mining Governance Indicator	MGI				NRGI; RAJIT
	Mining sector Transparency Indicator	MTI				NRGI; RAJIT
	Control of Corruption Indicator	CCI				NRGI, Transparency International
	Governance and Transparency Index	**GTI**	**9.52**	**39.5**	**20**	

References

Adiansyah, J. S., Rosano, M., Vink, S., & Keir, G. (2015). A framework for a sustainable approach to mine tailings management: Disposal strategies. *Journal of Cleaner Production, 108* (Part A): 1050–1062.

Aguna, C., & Kovacevic, M. (2010). Uncertainty and sensitivity analysis of the human development index. *Human Development Research Paper, 11.*

Ampella Mining Limited. (2013). Etude d'impact sur l'Environnement (EIE). Projet minier Batié-ouest, Burkina Faso. Final. Ouagadougou, Burkina Faso : Bureau National des Evaluations Environnementales, Ministère de l'Environnement et du Développement Durable, Burkina Faso.

Andersen, J. J., & Aslaksen, S. (2008). Constitutions and the Resource Curse. *Journal of Development Economics, 87*(2), 227–246.

Arezki, R., Gylfason, T., & Sy, A. (2012). Beyond the curse: Policies to harness the power of natural resources. *AfricaGrowth Agenda*, 12–14.

Auty, R. M. (2001). *Resource Abundance and Economic Development.* Oxford University Press.

Bai, C., Kusi-Sarpong, S., & Sarkis, J. (2017). An implementation path for green information technology systems in the ghanaian mining industry. *Journal of Cleaner Production, 164* (Supplement C), 1105–1123.

Ballet, J., Dubois, J.-L., & Mahieu, F.-R. (2012). La soutenabilité sociale du développement durable : de l'omission à l'émergence. *Mondes en développement, 156*(4): 89–110.

Bardhan, P. (1997). Corruption and development: A review of issues. *Journal of Economic Literature, 35*(3), 1320–1346.

Bayulken, B., & Huisingh, D. (2015). A literature review of historical trends and emerging theoretical approaches for developing sustainable cities (Part 1) *Journal of Cleaner Production,* 109 (Supplement C), 11–24. Special Issue: Toward a Regenerative Sustainability Paradigm for the Built Environment: From vision to reality.

Bini, L., M. Bellucci, & Giunta, F. (2018). Integrating sustainability in business model disclosure: evidence from the UK mining industry. *Journal of Cleaner Production, 171* (Supplement C), 1161–1170.

Bissa Gold, S. A. (2014). *Etude d'Impact sur l'Environnement du projet d'extension du permis d'exploitation aurifère de Bissa/Zandkom.* Ouagadougou, Burkina Faso: Bissa Gold SA.

Bolt, K., Markandya, A., Ruta, G., Lange, G.-M., Hamilton, K., Saeed Ordoubadi, M. et al. (2005). Where is the wealth of nations? Measuring capital for the 21st century. 34855. The World Bank.

Bouyssou, D. (1986). Second EURO summer institute some remarks on the notion of compensation in MCDM. *European Journal of Operational Research, 26*(1), 150–160.

Bouyssou, D., & Vansnick, J.-C. (1986). Noncompensatory and generalized noncompensatory preference structures. *Theory and Decision, 21*(3), 251–266.

Brand, D. A., Saisana, M., Rynn, L. A., Pennoni, F., & Lowenfels, A. B. (2007). Comparative analysis of alcohol control policies in 30 countries. *PLoS Med, 4*(4), e151.

Brollo, F., Nannicini, T., Perotti, R., & Tabellini, G. (2010). The political resource curse. Working Paper 15705. National Bureau of Economic Research.

Brundtland, G. H. (1987). World commission on environment and development. *Our Common Future.* Oxford: Oxford University Press.

Cadot, O., Carrère, C., & Strauss-Kahn, V. (2013a). Trade diversification, income, and growth: What do we know? *Journal of Economic Surveys, 27*(4), 790–812.

Cadot, O., Iacovone, L., Pierola, M. D., & Rauch, F. (2013b). Success and failure of African exporters. *Journal of Development Economics, 101*(March), 284–296.

Caiado, R. G. G., de Freitas Dias, R., Mattos, L. V., Quelhas, O. L. G. , & Leal Filho, W. (2017). Towards sustainable development through the perspective of eco-efficiency-A systematic literature review. *Journal of Cleaner Production, 165* (Supplement C), 890–904.

Carney, D. (1998). Sustainable rural livelihoods: What contribution can we make? Papers Presented at the Department for International Development's Natural Resources Advisers' Conference, July 1998. In *Sustainable Rural Livelihoods: What Contribution Can We Make? Papers Presented at the Department for International Development's Natural Resources Advisers' Conference, July 1998.* Department for International Development (DFID).

Chamaret, A. (2007). Une démarche top-down/bottom-up pour l'évaluation en termes multicritères et multi-acteurs des projets miniers dans l'optique du développement durable. Application sur les mines d'Uranium d'Arlit (Niger). Université de Versailles-Saint Quentin en Yvelines.

Chen, D., Ma, X., Hairong, M., & Li, P. (2010). The inequality of natural resources consumption and its relationship with the social development level based on the ecological footprint and the HDI. *Journal of Environmental Assessment Policy and Management, 12*(01), 69–86.

Cherchye, L., Moesen, W., Rogge, N., & Van Puyenbroeck, T. (2006). An introduction to 'Benefit of the Doubt' composite indicators. *Social Indicators Research, 82*(1), 111–145.

Cherchye, W., Moesen, N., Rogge, T. Van, Puyenbroeck, M., Saisana, A., Saltelli, R. Liska, et al. (2008). Creating composite indicators with DEA and robustness analysis: The case of the technology achievement index. *Journal of the Operational Research Society, 59*(2), 239–251.

Chinn, M. D., & Ito, H. (2008). A new measure of financial openness. *Journal of Comparative Policy Analysis: Research and Practice, 10*(3), 309–322.

Clément, M., Douai, A., & Gondard-Delcroix, C. (2012). Réflexions sur le concept de soutenabilité sociale dans le contexte des pays du Sud. *Mondes en développement, 156*(4): 7–18.

Code minier du Burkina Faso. (2015). Conseil national de la Transition, Burkina Faso.

Cominak. (2012). Bilan social 2012. Niamey, Niger: Areva Niger.

Costanza, R. (1992). *Ecological Economics: The Science and Management of Sustainability.* Columbia University Press.

Coste, J., Bouée, S., Ecosse, E., Leplège, A., & Pouchot, J. (2005). Methodological issues in determining the dimensionality of composite health measures using principal component analysis: Case illustration and suggestions for practice. *Quality of Life Research, 14*(3), 641–654.

Daly, H. E. (1990). Toward some operational principles of sustainable development. *Ecological Economics, 2*(1), 1–6.

Desai, M. (1991). Human development: concepts and measurement. *European Economic Review, 35*(2–3), 350–357.

Dialga, I. (2015). Du boom minier au Burkina Faso, opportunité de développement ou risques de péril pour des générations futures ? *Revue Cedres Etudes Sciences Economiques, Cedres, 59* (September), 27–47.

Dialga, I. (2018). A sustainability index of mining countries. *Journal of Cleaner Production, 179* (April), 278–291.

Dialga, I., & Le, T. H. G. (2017). Highlighting methodological limitations in the steps of composite indicators construction. *Social Indicators Research, 131*(2), 441–465.

Dubois, J.-L. (2009). The search for socially sustainable development conceptual and methodological issues. *Against Injustice: The New Economics of Amartya Sen.* Cambridge: Cambridge University Press, 275–294.

Endeavour mining-Avion Gold Burkina Sarl. (2014). "Etude d'impact environnemental, projet minier Houndé, Burkina Faso." Final. Ouagadougou, Burkina Faso: Bureau National des Evaluations Environnementales, Ministère de l'Environnement et du Développement Durable, Burkina Faso.

Field, F., Kirchain, R., & Clark, J. (2000). Life-cycle assessment and temporal distributions of emissions: Developing a fleet-based analysis. *Journal of Industrial Ecology, 4*(2), 71–91.

Frankel, J. A. (2010). The natural resource curse: A survey. Working Paper 15836. National Bureau of Economic Research.

Reed, M., Fraser, Evan D. G., Dougill, A. J., Mabee, W. E., & McAlpine, P. (2006). Bottom up and Top down: Analysis of participatory processes for sustainability indicator identification as a pathway to community empowerment and sustainable environmental management. *Journal of Environmental Management, 78*(2), 114–127.

Gamu, J., Le Billon, P., & Spiegel, S. (2015). Extractive industries and poverty: A review of recent findings and linkage mechanisms. *The Extractive Industries and Society, 2*(1), 162–176.

Gandhi, N. S., Thanki, S. J., & Thakkar, J. J. (2018). Ranking of drivers for integrated lean-green manufacturing for Indian manufacturing SMEs. *Journal of Cleaner Production, 171* (Supplement C), 675–689.

Gastauer, M., Silva, J. R., Junior, C. F. C., Ramos, S. J., Filho, P. W. M. S., Neto, A. E. F., et al. (2018). Mine land rehabilitation: modern ecological approaches for more sustainable mining. *Journal of Cleaner Production, 172*(January), 1409–1422.

Geiregat, C., & Yang, S. (2013). Des richesses trop abondantes? Fonds monétaire internationale. *Finance & Développement, 50*(3), 57.

Gelb, A. (2010). *Diversification de L'économie Des Pays Riches En Ressources Naturelles* (pp. 1–27). Washington DC: Fonds Monétaire International.

Graetz, G. (2014). Uranium mining and first peoples: The nuclear renaissance confronts historical legacies. *Journal of Cleaner Production, 84*(Supplement C), 339–47. Special Volume: The sustainability agenda of the minerals and energy supply and demand network: An integrative analysis of ecological, ethical, economic, and technological dimensions.

Gryphon Minerals Limited. (2014). Projet aurifère de Banfora, Burkina Faso, étude d'impact environnemental et social. Final. Ouagadougou, Burkina Faso: Bureau National des Evaluations Environnementales, Ministère de l'Environnement et du Développement Durable, Burkina Faso.

Hallstedt, S. I. (2017). Sustainability criteria and sustainability compliance index for decision support in product development. *Journal of Cleaner Production, 140*(Part 1), 251–66. Systematic Leadership towards Sustainability.

Hartwick, J. M. (1977). Intergenerational equity and the investing of rents from exhaustible resources. *The American Economic Review, 67*(5), 972–974.

Hartwick, J. M. (1990). Natural resources, National accounting and economic depreciation. *Journal of Public Economics, 43*(3), 291–304.

Hausmann, R., & Rigobon, R. (2003). An alternative interpretation of the 'Resource Curse': Theory and policy implications. Working Paper 9424. National Bureau of Economic Research.

Heravi, G., Fathi, M., & Faeghi, S. (2015). Evaluation of sustainability indicators of industrial buildings focused on petrochemical projects. *Journal of Cleaner Production, 109*(Supplement C), 92–107. Special Issue: Toward a Regenerative Sustainability Paradigm for the Built Environment: from vision to reality.

Hilson, G. (2010). 'Once a Miner, always a Miner': Poverty and livelihood diversification in Akwatia, Ghana. *Journal of Rural Studies, 26*(3), 296–307.

Hodge, R. A. (2014). Mining company performance and community conflict: Moving beyond a seeming Paradox. *Journal of Cleaner Production, 84*(Supplement C), 27–33. Special Volume: The sustainability agenda of the minerals and energy supply and demand network: an integrative analysis of ecological, ethical, economic, and technological dimensions.

Homma, T., & Saltelli, A. (1996). Importance measures in global sensitivity analysis of nonlinear models. *Reliability Engineering & System Safety, 52*(1), 1–17.

Hotelling, H. (1931). The economics of exhaustible resources. *Journal of Political Economy, 39* (2), 137–175.

Hugon, P. (2009). Le rôle des ressources naturelles dans les conflits armés africains. *Hérodote* n° 134 (3): 63–79.

IAM Gold Essakan SA. (2013). "Plan d'action réinstallation." Final. Ouagadougou, Burkina Faso: Bureau National des Evaluations Environnementales, Ministère de l'Environnement et du Développement Durable, Burkina Faso.

Irz, X., Lin, L., Thirtle, C., & Wiggins, S. (2001). Agricultural productivity growth and poverty alleviation. *Development Policy Review, 19*(4), 449–466.

Jensen, N., & Wantchekon, L. (2004). Resource wealth and political regimes in Africa. *Comparative Political Studies, 37*(7), 816–841.

Kalsoom, Q., & Khanam, A. (2017). Inquiry into sustainability issues by preservice teachers: A pedagogy to enhance sustainability consciousness. *Journal of Cleaner Production, 164* (Supplement C), 1301–1311.

Kanyimba, A. T., Richter, B. W., & Raath, S. P. (2014). The effectiveness of an environmental management system in selected South African primary schools. *Journal of Cleaner Production, 66*(March), 479–488.

Kaufmann, D., Kraay, A., & Mastruzzi, M. (2009). Governance matters VIII: Aggregate and individual indicators, 1996–2008. *World Bank Policy Research Paper*, no. 4978.

Kiaka Gold Sarl. (2014). Projet aurifère Kiaka, Burkina Faso, Etude d'impact environnemental et social, rapport final. Final. Ouagadougou, Burkina Faso: Kiaka Gold Sarl.

Kitula, A. G. N. (2006). The environmental and socio-economic impacts of mining on local livelihoods in Tanzania: A case study of Geita district. *Journal of Cleaner Production, 14* (3–4), 405–414. Improving Environmental, Economic and Ethical Performance in the Mining Industry. Part 1. Environmental Management and Sustainable Development Improving Environmental, Economic and Ethical Performance in the Mining Industry. Part 1. Environmental Management and Sustainable Development.

Kotsadam, A., & Tolonen, A. (2016). African mining, gender, and local employment. *World Development, 83*(Supplement C), 325–339.

Kusi-Sarpong, S., Sarkis, J., & Wang, X. (2016). Assessing green supply chain practices in the Ghanaian mining industry: A framework and evaluation. *International Journal of Production Economics, 181*(Part B), 325–341. Recent Development of Sustainable Consumption And Production In Emerging Markets.

Lagos, G., Peters, D., Videla, A., & Jara, J. J. (2018). The effect of mine aging on the evolution of environmental footprint indicators in the Chilean copper mining industry 2001–2015. *Journal of Cleaner Production, 174*(Supplement C), 389–400.

Leduc, G. A., & Raymond, M. (2000). L'évaluation des impacts environnementaux: Un outil d'aide à la décision. *VertigO - la revue électronique en sciences de l'environnement*, 403.

Levine, S. H., Gloria, T. P., & Romanoff, E. (2007). A dynamic model for determining the temporal distribution of environmental Burden. *Journal of Industrial Ecology, 11*(4), 39–49.

Lu, Y., Geng, Y., Liu, Z., Cote, R., & Yu, X. (2017). Measuring sustainability at the community level: An overview of China's indicator system on national demonstration sustainable communities. *Journal of Cleaner Production, 143*(Supplement C), 326–335.

Marin, T., Seccatore, J., De Tomi, G., & Veiga, M. (2016). Economic feasibility of responsible small-scale gold mining. *Journal of Cleaner Production, 129*(Supplement C), 531–536.

Midas Gold Sarl. (2013). *Etude d'impact environnemental et social: unité de traitement semi-mécanisé des haldes et stériles*. Ouagadougou, Burkina Faso: Midas Gold Sarl.

Munda, G. (2005). 'Measuring Sustainability': A multi-criterion framework. *Environment, Development and Sustainability, 7*(1), 117–134.

Munda, G., & Nardo, M. (2003). *On the methodological foundations of composite indicators used for ranking countries*. Ispra, Italy: Joint Research Centre of the European Communities.

Munda, G., & Nardo, M. (2009). Noncompensatory/Nonlinear composite indicators for ranking countries: A defensible setting. *Applied Economics, 41*(12), 1513–1523.

Nathan, H. S. K., Mishra, S., & Reddy, B. S. (2012). An alternative approach to measure HDI. Working Paper of IGIDR 1. Mumbai, India: Indira Gandhi Institute of Development Research, Mumbai.

Northey, S. A., Mudd, G. M., Saarivuori, E., Wessman-Jääskeläinen, H. & Haque, N. (2016). Water footprinting and mining: Where are the limitations and opportunities?" *Journal of Cleaner Production, 135*(Supplement C), 1098–1116.

O'Connor, M., & Spangenberg, J. H. (2008). A methodology for CSR reporting: Assuring a representative diversity of indicators across stakeholders, scales, sites and performance issues. *Journal of Cleaner Production, 16*(13), 1399–1415.

OECD & JRC. (2008). *Handbook on Constructing Composite Indicators: Methodology and User Guide*. OECD Publishing.

Oxfam. (2013). "Niger: A qui profite l'uranium? L'enjeu de la renégociation des contrats miniers d'Areva." Oxfam International.

Papyrakis, E., & Gerlagh, R. (2004). The resource curse hypothesis and its transmission channels. *Journal of Comparative Economics, 32*(1), 181–193.

Pearce, D. W., Warford, J. J., & others. (1993). *World without End: Economics, Environment, and Sustainable Development*. Oxford University Press.

Perišić, A. (2015). Data-driven weights and restrictions in the construction of composite indicators. *Croatian Operational Research Review, 6*(1), 29–42.

Piccinno, F,, Hischier, R., Seeger, S., Som, C. (2018). Predicting the environmental impact of a future nanocellulose production at industrial scale: Application of the life cycle assessment scale-up framework. *Journal of Cleaner Production, 174*(Supplement C), 283–295.

Plan African Minerals LTD. (2014). "Etude d'impact environnemental et social du projet d'exploitation du manganèse de Tambao." Final. Ouagadougou, Burkina Faso : Bureau National des Evaluations Environnementales, Ministère de l'Environnement et du Développement Durable, Burkina Faso.

Rebai, S., Azaiez, M. N., Saidane, D. (2016). A multi-attribute utility model for generating a sustainability index in the banking sector. *Journal of Cleaner Production, 113*(Supplement C), 835–849.

Ryoo, S. Y., & Koo, C. (2013). Green practices-IS alignment and environmental performance: The mediating effects of coordination. *Information Systems Frontiers, 15*(5), 799–814.

Saisana, M., & Saltelli, A. (2010). *Uncertainty and Sensitivity Analysis of the 2010 Environmental Performance Index*. OPOCE.

Salisu Barau, A., Stringer, L. C., & Adamu, A. U. (2016). Environmental ethics and future oriented transformation to sustainability in Sub-Saharan Africa. *Journal of Cleaner Production, 135*(Supplement C), 1539–1547.

Sba-Ecosys-Cedres. (2011). Analyse économique du secteur des mines : liens pauvreté et environnement. Rapport final. Ouagadougou, Burkina Faso : Ministère de l'Environnement et du Développement durable.

SEMAFO Burkina SA. (2013). Etude d'impact environnemental et social du projet F1 : extension du permis de Mana. Ouagadougou, Burkina Faso : SEMAFO SA, Société d'Exploration Minière en Afrique de l'Ouest.

SOMAIR-Société des Mines de l'Aïr. (2012). Rapport Environnemental, Social et Sociétal 2011–2012. Annuel. Niamey, Niger : Areva Niger.

SOMINA-Société des Mines d'Azélik. (2014). Rapport Environnemental, Social et Sociétal 2013–2014. Annuel. Niamey, Niger: SOMINA.

Sullivan, K., Thomas, S., & Rosano, M. (2018). Using industrial ecology and strategic management concepts to pursue the sustainable development goals. *Journal of Cleaner Production, 174*(February), 237–246.

Tahmasebi, M., Feike, T., Soltani, A., Ramroudi, M., & Ha, N. (2018). Trade-off between productivity and environmental sustainability in Irrigated vs. Rainfed wheat production in Iran. *Journal of Cleaner Production, 174*(February), 367–379.

Tarantola, S., & Vertesy, D. (2012). *Composite indicators of research excellence.* Institute for the Protection and Security of the Citizen: EUR-Scientific and Technical Research Reports.

Turcu, C. (2013). Re-thinking sustainability indicators: Local perspectives of urban sustainability. *Journal of Environmental Planning and Management, 56*(5), 695–719.

Valenzuela-Venegas, G., Cristian Salgado, J., & Díaz-Alvarado, F. A. (2016). Sustainability indicators for the assessment of eco-industrial parks: Classification and criteria for selection. *Journal of Cleaner Production, 133*(Supplement C), 99–116.

Van der Ploeg, F. (2011). Natural resources: Curse or blessing? *Journal of Economic Literature, 49*(2), 366–420.

Wang, M.-X., Zhao, H.-H., Cui, J.-X., Fan, D., Lv, B., Wang, G., et al. (2018). Evaluating green development level of nine cities within the Pearl River Delta, China. *Journal of Cleaner Production, 174*(February), 315–323.

Yakovleva, N., Kotilainen, J., & Toivakka, M. (2017). Reflections on the opportunities for mining companies to contribute to the united nations sustainable development goals in Sub–Saharan Africa. *The Extractive Industries and Society, 4*(3), 426–433.

Zeijl-Rozema, A., & Martens, P. (2010). An adaptive indicator framework for monitoring regional sustainable development: A case study of the INSURE project in Limburg, The Netherlands. *Sustainability: Science, Practice, & Policy, 6*(1).

Zeleny, M. (1982). Multi criteria decision making. *TIMS Studies in Manage*, 31–57.

Zhang, Q., Yue, D., Fang, M., Yu, Q., Huang, Y., Su, K., Ma, H., & Wang, Y. (2018). Study on sustainability of land resources in Dengkou county based on emergy analysis. *Journal of Cleaner Production, 171*(Supplement C), 580–591.

Zhou, P., & Ang, B. W. (2008). Comparing MCDA aggregation methods in constructing composite indicators using the Shannon-Spearman measure. *Social Indicators Research, 94*(1), 83–96.

Zhou, P., Ang, B. W., & Poh, K. L. (2006). Comparing aggregating methods for constructing the composite environmental index: An objective measure. *Ecological Economics, 59*(3), 305–311.

Application of Indicators in Transport Planning: Insight from India

Deepty Jain

Abstract In India, transport sector consumes nearly 9% of the total oil demand contributing to approximately 7.5% of the total greenhouse gas emissions. Accounting for fuel consumption and emissions throughout lifecycle of transport infrastructure, services and vehicles will further expedite the numbers. National urban transport policy (NUTP) emphasize on take-up of strategies and policies to encourage the use of non-motorized transport and public transport in Indian cities. Appropriate adaptation of NUTP shall help cities in reducing environmental impacts of transport sector. This requires appropriate planning and management of infrastructure and services that reduce impacts both in short and long term. The approach relies on the use of indicators to guide the process of defining strategies and monitoring of the progress made. In past two decades, urban transport in India has seen transition from provision of infrastructure to provision of services. Various plans and studies have been taken up to support urban local bodies. The chapter studies the transition in urban transport policies, approaches for planning urban transport and application of indicators in achieving the laid agenda in past two decades in India. In India, mobility planning is divided into three phases—prior to the issue of NUTP in 2006, after the launch of JnNURM (2006–2014), and after the revised toolkit (R-toolkit) for preparing Comprehensive Mobility Plan (CMP) was issued (beyond 2014). R-toolkit provides an extensive list of indicators to measure transport infrastructure availability and impact of transport system in existing condition and for alternate scenarios on society and environment. As per the R-toolkit, the CMP should propose strategies and prioritize them based on the scenario analysis. However, the review of CMP for Udaipur and Salem shows that the indicators related to measuring health, safety, and security were not measured. This is due to the complexity involved in measuring these indicators and lack of R-toolkit in defining method for measuring them. There is a need to develop methods and tools for measuring these indicators that can be adopted by the local authorities and/or the consultants preparing the CMP. The objectives laid in NUTP

D. Jain (✉)
Department of Energy and Environment, TERI School of Advanced Studies,
Plot no. 10, Vasant Kunj Institutional Area, New Delhi 110070, India
e-mail: archikooldeepty@gmail.com

© Springer Nature Singapore Pte Ltd. 2019
S. S. Muthu (ed.), *Development and Quantification of Sustainability Indicators*,
Environmental Footprints and Eco-design of Products and Processes,
https://doi.org/10.1007/978-981-13-2556-4_3

can be achieved by CMP by integrating CMP in the existing legal and institutional framework and developing monitoring mechanisms to assess implementation of CMP.

Keywords Mobility plan · Indicators · Planning approaches · Sustainable mobility · Indian cities · Sustainable mobility

1 Introduction

Globally, focus of transport planning has shifted from increasing mobility of vehicles to increasing accessibility and safety, strengthening public transport systems, encouraging mobility by non-motorized transport, enhancing social equity and achieving environmental sustainability in past two decades. This has changed the methodology adopted for preparing transport plans that requires application of indicators and informed decision-making. Although take-up of these approaches was slow in the first decade of 2000s, however a greater momentum has been realized in take-up of transport planning for people in the second decade all across the world.

Similarly, in India too, a paradigm shift in transport planning is observed. It starts with 74th Amendment Act (74thAA) of Indian constitution in 2004 that deliberated powers and responsibilities to urban local bodies (ULBs) for preparing and implementing plans and schemes that are in favour of economic development and social justice. The amendment gave greater flexibility to urban local bodies to prepare development plans, build capacity and disburse resources. Later in 2006, National Urban Transport Policy (NUTP) was adopted by Ministry of Urban Development (MoUD) with a focus on providing better services and mobility to people as compared to the vehicles. Following this, Jawaharlal Nehru National Urban Missions (JnNURM) was launched to provide funding to urban local bodies for upgrading basic infrastructure in the cities. To enable the process, JnNURM required cities to prepare City Development Plan (CDP) to identify issues and challenges and propose strategies panning across all the sectors. ULBs were also required to prepare Comprehensive Mobility Plans (CMPs) to define issues and strategies in alignment to NUTP, the toolkit for which was issued in 2008.

Aim of the chapter is to understand the role of indicators in transport planning to achieve the goals of sustainable mobility in global and Indian context. The chapter specifically deliberates on the paradigm shift in methods used to prepare transport plans in India and type of indicators used in various planning documents. It also deliberates on the adherence of city-level transport plans with national visions and goals and application of indicators in practice.

The chapter is further divided into four parts. In Sect. 2, indicators are defined. Third section discusses different definitions of sustainable transport adopted by various countries. Later, the parameters defined in the national visions of various countries are compared to understand how countries deliberate more or less on one or more pillars of sustainable transport. This therefore helps in identifying priority agenda for the countries. In the fourth section, mobility or transport planning approaches adopted in developed (Europe and North America) and developing (Africa and Latin America) regions are discussed. This section also compares city-level transport plans and indicators used in practice with regard to achieving country-specific goals taking examples from developed countries. In the last section, various transport planning approaches adopted over time in India are discussed. This section also discusses the indicators proposed in the guidelines issued by national government for preparing various types of planning documents. Later in the section, application of indicators in practice taking two examples of city-level mobility plan is discussed. In conclusion, the need for selection and use of appropriate indicators in transport plans is stressed to achieve the objectives of sustainable development.

2 Indicators Based Approach

Transport planning can be broadly classified into two approaches that is 'predict and provide' approach and scenario-based approach. Former relates to estimating potential travel demand for current and future years and providing solutions such as road widening schemes and flyovers to ensure smooth flow of traffic. This approach does not involve evaluation of projects or strategies with regard to environmental and societal impacts. As compared to this, scenario-based approach involves evaluating alternate scenarios or strategies of infrastructure improvement and urban development against the defined targets. This requires application of indicators that are in alignment to the laid objectives and help in reflecting contextual issues and challenges. Based on the evaluation, best scenario is selected and priority projects are identified. Therefore, the later approach helps in informed decision-making that is useful for managing travel demand and minimizing negative impacts of transport sector on environment and society in both short and long term.

Indicators are measurable parameters that help in evaluating current situation and alternate scenarios. Organization for Economic Cooperation and Development Countries (OECD) defined indicators as statistical measures of social, environmental and economic sustainability (Haghshenas and Vaziri 2012). Various organizations are using transport indicators to evaluate progress, projects, and policies towards set goals and objectives. Indicators help in evaluating, simplifying, study trends, communicate issues and compare across places and situations (Boyko et al. 2012; DETR 2000b; Toth-Szabo and Várhelyi 2012). A set of appropriate indicators allow decision makers to monitor status and understand consequences of the actions and inactions (cited in Jain and Tiwari 2017b).

Various institutes and authorities have developed sustainable mobility indicators for efficient planning. Even though consensus on meeting the 'triple bottom line' exists, i.e. environmental, social and economic sustainability; yet different indicator sets have been used to evaluate transport systems (Miranda and Silva 2012; Richardson 2005). It is required to select limited set of indicators that provides holistic understanding of the system to identify relevant issues and select policies for meeting the planning targets (Jain and Tiwari 2017b).

The selection of indicators and application depends on the goals identified by the respective organizations. In general, organizations identify goals and visions to be achieved in the horizon year and then translate these goals into objectives. For each objective, indicators are selected to define benchmarks and specific targets sometimes also termed as performance indicators. Therefore, it is likely to have more number of indicators for the goal that is of higher priority for the organization.

2.1 What are Good Indicators?

Indicators are used to enable quantification of objectives, defining benchmarks and analyse trends such that issues can be identified and strategies can be prioritized. Therefore, an indicator should be quantifiable, measurable, transparent and interpretable. Indicators also enable data collection, but for measuring indicators if the data acquisition is difficult then measuring such indicator over a temporal scale can pose limitations. Therefore, selection of indicators should also be based upon data availability and data collection procedure to be adopted for measuring it (Dale and Beyeler 2001; Gilbert et al. 2002). The selected indicator or indicators should also allow comparison across regions and timescale. Since indicators are also used to identify best scenarios against the defined benchmarks and to prioritize projects, the indicators that can be forecasted are more suitable specifically for preparing development plans (Castillo and Pitfield 2010; Moussiopoulos et al. 2010).

One indicator can address one or two planning objectives or more than one indicator may address an objective. Therefore, it is essential to understand that there is a set of indicators for measuring multiple aspects of the system under consideration. The indicators used within a defined set should not conflict each other but should be coherent to provide policy direction (Jain and Tiwari 2017b; Litman 2009; Toth-Szabo and Várhelyi 2012). An example of two conflicting targets can be reducing greenhouse gas emissions (measured as CO_2 equivalent emissions from transport) and reducing congestion (measured as volume–capacity ratio). While, one may argue that both lead to sustainable development that is by reducing vehicular congestion on road, we can reduce greenhouse gas emissions. However, referring to demand–supply curve, targeting congestion is likely to result in overall increase in personal vehicle use in long term therefore negating the impact of reduced congestion on emission levels. The selected indicators should provide unambiguous, specific information that can be used for decision-making to achieve the defined goals.

3 Sustainable Transport Definition and Visions

Policies, targets and visions defined by the national government influence transport planning process and outcomes at local level. Clarity in the definition of visions and sustainable mobility has shown to have impact on the quality of the plans prepared (May et al. 2012; Rosqvist and Wennberg 2012).

Sustainable transport and planning visions have been defined differently by various organizations. Such examples are

- A sustainable transportation system must be safe, efficient in providing accessibility and mobility, and in enhancing economic productivity, without impacting the natural environment negatively—all in a manner that is equitable to those who use and are affected (either directly or indirectly) by the system (Amekudzi et al. 2009).
- The definition adopted by Centre for Sustainable Transportation (CST), Canada (2002) states that a transport system is sustainable if it meets the need of individuals with minimum mobility, especially motorized mobility and use minimum non-renewable resources. The definition encompasses the 17 elements as follows:

    - "Allows the basic access of individuals and societies to be met safely and in a manner consistent with human and ecosystem health and with equity within and between generations;
    - Is affordable, operates efficiently, offers choice of transport mode and support a vibrant economy;
    - Limits emissions and waste within the planet's ability to absorb them, limits consumption of renewable resources to the sustainable yield level, minimizes consumption of non-renewable resources, reuses and recycles its components and minimizes the impacts on the use of land and the production of noise".

- Definition adopted by European Council states that a 'sustainable transport system meets society's economic, social and environmental needs whille minimizing its undesirable impacts on the economy, society and the environment' (Bührmann et al. 2011).
- In India, NUTP (2014) states the objective of ensuring safe, affordable, quick, comfortable, reliable and sustainable access for the growing number of city residents to jobs, education, recreation and such other needs within our cities for transport planning.

and so on ...

Different organizations and country have adopted various visions for planning transport system that in common align to environmental, economic and social sustainability; however, the depth in which each component has been dealt differs. There is also a difference in definitions adopted between developed and developing countries.

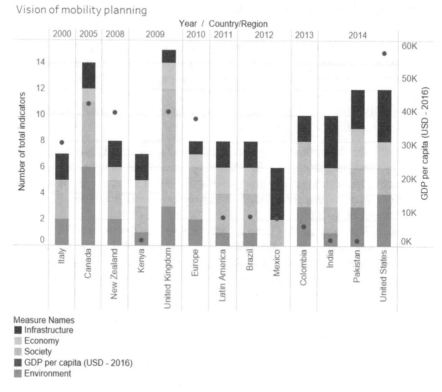

Fig. 1 Vision parameters for sustainable transport in different Countries Source: Author data compiled from various sources: (CST Canada 2005; Department for Transport 2009; Federal Highway Administration USA 2012; Hidalgo and Huizenga 2013; Leo et al. 2017; Lopez-Lambas et al. 2013; Ministry of Transport New Zealand 2008; Ministry of Urban Development 2014; Republic of Kenya 2009; Rupprecht Consult 2011; The President of the Republic 2012; Tyler et al. 2013)

Figure 1 shows the parameters defined in national visions for attaining the goals of sustainable transport in various countries along with the current gross domestic product per capita (USD) for 2016. The parameters are classified into four groups of economic, environmental and social sustainability including infrastructure provision. However, due to the complexity of the system, one parameter can be related to multiple aspects as well.

As compared to other countries, maximum number of parameters has been included in the national vision of Canada and UK for defining sustainable mobility. In Canada, approximately 43% of the parameters are related to environmental sustainability that includes conserving natural environment, reducing greenhouse gas emissions, achieving energy efficiency, reducing dependency on non-renewable resources of energy and such others. Whereas, in UK 60% of the parameters are associated with achieving social sustainability. This includes achieving equity and equality, provision of accessible, affordable, acceptable, safe, and secure transport that helps in improving quality of life and health of the citizens. In the USA, equal emphasis is on attaining environmental sustainability and infrastructure provision. In low and low-middle income countries, parameters are majorly related to economy that includes, achieving economic growth and finance management and infrastructure provision, i.e. reliable, smart, and efficient system that helps in reducing travel time.

4 Mobility Planning and Role of Indicators in Global Context

Transport sector for the first time was identified as crucial sector for achieving sustainable development in the year 1992 by the United Nations (United Nations Sustainable Development 1992). The action plan emphasized on developing cost-effective, efficient, less polluting and safer transport systems for urban settings. To achieve this, the agenda called for collecting, analysing and exchanging relevant information related to the impact of transport on energy consumption, emissions and safety. It identified integrating land use and transport along with development of technologies to reduce impact of transport sector on environment. In the year 2005, Kyoto Protocol adopted by United Nation Framework Convention on Climate Change came into action. It specified development of national and regional programmes that should include measures to mitigate and facilitate adaptation to climate change with focus on transport sector as one of the major sectors affecting climate change (UNFCCC 1997). Following this, various countries formulated national strategies for development of sustainable transport in cities, wherein the role of mobility plans was identified to guide the process of selection and implementation of strategies that can help in achieving the goals. Sustainable transport plans define long-term vision for the cities and strategies to achieve the same. The strategies are identified by using indicators and scenario-based approach in consensus with all the parties of interest.

4.1 Mobility Planning

4.1.1 Europe

European communities defined regional strategy for developing sustainable urban mobility plans in 2006 (Commission of the European Communities 2007). Prior to this, Plans de De´placements Urbains (PDU) in France were developed since 1980s (Lopez-Lambas et al. 2013) and in UK, first guideline for preparing local transport plans (LTP) was issued in 2000 (DETR 2000a). Figure 2 shows the timeline when

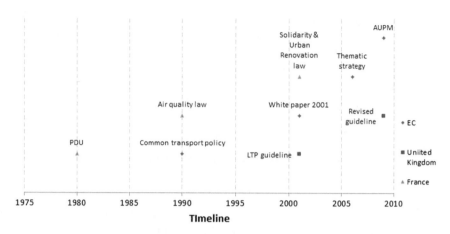

Fig. 2 Mobility planning in European Countries

decisions related to adoption of mobility planning approach were taken in various parts of Europe.

4.1.2 Plans de De'Placements Urbains (PDU) in France

PDU had been in existence since 1980s that aimed to organize urban transport to encourage rational use of car and integrate non-motorized transport and public transport for providing alternate equivalent choices in urban areas. In 1996, air quality law mandated urban areas (population > 100,000) to prepare PDU and Solidarity and Urban Renovation law (2000) placed it within the hierarchy of other planning documents. These policies gradually transformed PDU into a central planning document (specify timings of decision and meet objectives of sustainable mobility) providing it a more statuary status. Lopez-Lambas et al. (2013) identified two key features of PDU—stakeholder participation and coordination between mobility and urban development strategies—that made it different from other mobility planning approaches.

4.1.3 Local Transport Plan in UK

The Department of Environment, Transport and Regions (DETR), UK in 2001 published guidelines for preparing LTP to improve safety and accessibility for all transport users and protect and enhance the built and natural environment. The guideline defined two key steps for preparing LTP—anticipation of the future travel demand and use of the measurable indicators to choose a strategy over the other alternatives (DETR 2000a). As per the guidelines, the local authorities are required to select indicators that should reflect the national targets. If the local authorities fail to account for the indicators identified in the national strategy, then they were required to provide appropriate reasons for not including them. The too flexible approach resulted in the failure of achieving the aims of sustainable mobility planning (House of Commons 2006; May et al. 2008).

In 2009, the Department for Transport, London, issued new guidelines for preparing LTP considering the improvements over the previously issued set of guidelines in 2000 and 2004. It described steps related to the identification of the problems, defining objectives, appraisal and selection of transport options (Department for Transport 2009). The guidelines provided a list of both mandatory and optional indicators to evaluate measures and the choice of using optional indicators was left to the local authorities (Department for Transport 2009) making the process more adaptive to the local context (Lopez-Lambas et al. 2013).

4.1.4 Sustainable Urban Mobility Plans in European Union

'White Paper' submitted in 2001 highlighted that the Common Transport Policy adopted in 2001 resulted in increased personal mobility and negative impact on human health and environment (Commission of the European Communities 2001). The 'White Paper 2001' called for attention to develop a comprehensive strategy comprising of land use policies considering both spatial and temporal distribution of activities along with fiscal and transport research-based policies.

In 2006, the European Commission (EC) adopted a thematic strategy identifying urban mobility as one of the major contributors to the degrading environment and challenges faced by most of the urban areas in the European Union. It called for the need to prepare Sustainable Urban Mobility Plans (SUMP) (Commission of the European Communities 2007) following which the first guideline for preparing SUMP was issued (Rupprecht Consult 2011). In the year 2009, the commission adopted Action Plan on Urban Mobility that defined the role of European Commission in supporting and enabling the efforts at local level by building capacity of the local authorities, providing guidelines, promoting exchange of the best practices and identifying benchmarks (Commission of the European Communities 2009). The revised guidelines issued in 2009 specified three key steps for preparing SUMP. These were—analyzing the existing situation, developing scenarios and selecting effective measures. The guideline provides an overview of the transport models and the basis for selection of the measurable indicators (Bührmann et al. 2011).

Countries like Italy, Denmark, Spain have developed national frameworks for preparing SUMP. Cities like Burgas in Bulgaria, Gewest Brussels in Belgium, Czestochowa in Poland and Budapest in Hungary have prepared SUMP even though the respective national guidelines have not mandated it (European Commission 2013).

The municipal corporation of Opava (Czech Republic) prepared SUMP in 2015 with a vision to meet mobility needs of the citizens in a sustainable way. The plan was prepared in three phases—strategic phase, analytical phase and proposal phase.

In the strategic phase objectives, indicators and likely strategies are identified and analytical phase included analysis of the existing transport system and future development patterns. The proposal phase consisted of development of an integrated package of measures (Eltis 2015).

In 2010, the local authority of Budapest, Hungary, decided to introduce a new management and organizational structure for transport. The plan then prepared included social cost-benefit analysis and multi-criteria analysis for selecting measures (Bohler-Baedeker et al. 2015). In 2013, the planning process was considered for revision to align development strategy with SUMP approach.

4.1.5 North America

In USA, national-level strategic interventions for preparing mobility or transport plans had been lacking until 2012. Although transport as a subject of importance has been part of various strategies like achieving sustainability, improving air quality and addressing climate change, however specific national level programme to achieve sustainable transport was not defined (Zhou 2012). United States Department of Transport (USDOT) (2006) adopted a strategic plan to provide fast, safe, efficient, and convenient transportation at the lowest cost that should also encourage efficient use and conservation of the resources. The strategic plan provided a list of indicators and targets to be achieved along with a list of strategies to be adopted for achieving the defined targets. However, it did not mandate local authorities to prepare mobility or transport plans in alignment to the national strategies. In 2012, transportation bill 'moving ahead for progress in twenty-first century' was adopted. It stated that urbanized areas with more than 50,000 population need to develop transport plans considering all modes of transport for promoting safe and efficient transportation systems to serve the needs of people and freight and foster economic growth and development within and between States and urbanized areas. The plan should help in reducing fuel consumption and air pollution. Transport planning at local level is the responsibility of Metropolitan Planning Organization (MPO), state Department of Transportation (state DOT), and transit operators (FHA and FTA 2015) in USA. State DOT prepares long-range statewide transportation plan and is responsible for development and maintenance of state infrastructure.

Sustainable transport planning initiated in 1997 in Canada, when, Transport Canada, the responsible federal agency defined strategies to develop sustainable transport in the country to address country-specific challenges and define strategies for the same. This was revised and updated thrice before a clear definition of sustainable transport was issued in 2005. Similar to the USA, Canada federal government did not mandate transport planning at local level. However, provincial planning acts mandated municipal governments to prepare land use plans within which transport plan shall be embedded. The land use plan prepared by the local authorities should adhere to the defined provincial principles and standards.

National government controls over the quality of the plans and ensures its adherence with national strategy through fiscal regulations. Transport-related projects are identified using cost and benefit analysis within which parameters like economic value of travel time, emissions and accidents are included (Ottawa City Services 2013). Vancouver in 2014 issued transportation plan 2040 that aligns with City's Greenest City 2020 Action Plan aiming to encourage maximum mobility by non-motorized transport, reduce travelled distance and eliminate fatalities related to transport. The plan also propose policies and strategies to track progress in the identified indicators over the years.

4.1.6 Latin America

There had been a growth in take-up of sustainable transport policy and planning in Latin America in last few decades. The successful examples of implementing Bus Rapid Transit Systems are from the cities of Latin America. In the year 2011, nine countries from South America including Mexico adopted Bogotá Declaration on sustainable transport objectives. The declaration laid 23 goals spreading over four strategies, i.e. to reduce dependency on motorized travel and reduce travel distances; to achieve safe, efficient and environmentally friendly transport system; to improve technology and management of transport services and adopt cross-cutting strategies (Hidalgo and Huizenga 2013). After this only, various national governments like that of Mexico, Colombia and Brazil identified urban mobility as essential development strategy and incorporated it in the federal system framework. The national transport strategy 2030 adopted in Colombia aims to improve accessibility that enables good quality of life, reduce energy consumption and emissions and ensure sustainable growth and development. The strategy specifies development of city-specific plans and policies in alignment to the national policy (Tyler et al. 2013).

In Brazil, the discussion for traffic management initiated in 1997 when the powers for local traffic planning and management were decentralized from regional to local level (Vasconcellos 2018). Later in 2000, City Statute law identified the importance of approach for integrated land use and transport planning. However, the law was not adopted by the cities with population less than 100,000 accounting to inadequate resources. Following this, urban mobility law in 2011 mandated cities with population more than 20,000 to prepare mobility plans to be able to receive funding from federal government. The law emphasized on achieving efficiency, equity and sustainability in transport sector by reducing or minimizing the unnecessary use of car. The Municipal Corporation of Belo Horizonte (Brazil) adopted an urban mobility plan in 2010 to ensure road safety, promote improvements in facilities and services and reduce impacts on the environment (Bohler-Baedeker et al. 2015). The plan discusses two long terms and one intermediate scenario for 2020 and 2014, respectively. Later in 2012, it was adopted as urban mobility master plan to conform to the Federal Law. The funding received from the national government was used to set up an observatory to monitor the implementation of the mobility plan.

4.1.7 Africa

Countries in Africa lacked in overall transport infrastructure development. Local and regional authorities had limited resources and institutional capacity to conduct studies, identify need, prepare plans and implement them. Lack of appropriate governance structure and national policy framework resulted in increasing concerns with regard to urban mobility and its impact on environment and human well-being. EuroMed transport project initiated in 2005 with adoption of regional transport action plan for Mediterranean region 2007–2013. One of the aims of the initiative is to enable reforms for adopting SUMP approach in the region (Euromed Transport Forum 2013).

Republic of Kenya in 2009 prepared an integrated national transport policy with an aim to provide efficient, cost-effective, safe, secure and integrated transport system to attain development in sustainable way. The policy provides a framework for better management of transport sector through reforms in governance, regulatory framework and integrated system based approach (Republic of Kenya 2009). It requires local authorities to develop integrated land use transport plans in consultation with other relevant authorities. The Integrated Urban Development Master Plan for Nairobi prepared in the year 2014 consist a section on urban transport development plan. The plan provides a list of priority projects like freight terminal development, road widening, staged office hours and improvement of non-motorized transport infrastructure on the specific corridors. The selection of priority projects was based on the estimated existing and future travel demand and scenario-based analysis. Impacts of developed scenarios were estimated on modal share, average traffic speeds and volume–capacity ratio (JICA 2014).

The large size cities of Morocco like Casablanca and Marrakech are adopting transport plans to improve efficiency and supply of transport infrastructure and contribute to environmental and social sustainability (El-Geneidy et al. 2013). This is taken up by both plan-led and vision-led approaches and integrating stakeholder participation at various stages of planning. Due to the lack of appropriate indicators and method for evaluation of the transport options, the objectives of sustainable mobility have not been achieved (Asmaa et al. 2012).

The countries of African continent have also adopted Agenda 2063 in 2013 with an aim to build an integrated, prosperous and peaceful Africa. It identifies the need to increase accessibility to transportation services as a priority area. The first 10-year implementation plan requires development and implementation of policies for better urban planning, land tenure, use and management systems; regulatory framework and expansion of infrastructure to provide affordable access to the transport services and facilitate development of urban mass transit systems through alternate financing mechanisms (African Union 2015).

4.2   Transport Plans and Indicators in Global Context

As discussed, organizations in different regions of the world have defined vision for sustainable transport adhering to the three pillars of sustainability; however, the detail with which each pillar has been addressed varies. This is to say that, some organizations lay more stress on economic sustainability while for some others priority is to attain social sustainability. Mapping of indicators with vision and goals can help in identifying how different organizations place importance to one or more goals as compared to the whole set. This will also help in understanding role of indicators in transport planning in practice to attain the goals defined in national policy document.

In this section, indicators used in transport plans are linked with the respective national visions and goals. For the purpose, examples of three mobility plans from developed nation, one each from USA, Canada and UK are presented. The selection of these mobility plans is arbitrary and largely based on ease with which these mobility plans were accessible in English language on public websites or domains.

Table 1 presents the indicators used in transport plans of New York (USA), Vancouver (Canada) and London (UK) against the defined vision for transport planning at national level for the three countries. The shaded cell in the table shows the presence of vision parameter in the national policy document and the dot in the cell shows the presence of indicator to measure and monitor the parameter in the transport plan. As shown in Fig. 1, the three countries have identified different number of parameters in national vision and policy document. In Canada, for example, equal number of parameters are identified for environmental and social sustainability. As compared to this in USA, equal thrust is upon achieving environmental sustainability and provision of infrastructure. In UK, focus is on achieving social sustainability followed by environmental aspects. However, different interpretation is drawn when the three transport plans belonging to each of the three countries are compared.

New York Transportation plan provides performance measure for traffic management measures only such as travel time, vehicle-kilometre, vehicle and person-hours of delay and vehicle kilometres. Specific indicators related to environmental and social sustainability are overall missing from the plan. Vancouver transportation plan for 2040 emphasize on three targets or goals, i.e. increase non-motorized and public transport share, reduce fatalities and reduce distance. In the plan, various policies have been identified like alternate fuel use or change in vehicle technology; however, targets or specific indicators for these aspects have not been identified in the plan. Mayors transport strategy for London (2018) specifies indicators and targets for the identified parameters in LTP guideline except for affordability, acceptability and climate resilience. Although the plan lays down short- and long-term strategies to achieve these goals, however specific targets and indicator to monitor the progress is missing.

The three countries UK, USA and Canada have high number of parameters considered in the national policy document. The analysis shows that London

Table 1 Linking indicators and vision for mobility planning

Agenda or parameters	Vision	Specific indicators	United States New York 2040	Canada Vancouver 2040	United Kingdom London 2018
Economy	Economic growth	New jobs from transport			•
	Cost effective				
	Finance				
Environment	GHG emissions	Emissions from vehicles			•
	Natural environment	Percentage green area			•
		No. of trees on street			•
	Energy efficiency				
	Fuel type				
	Land consumption and waste				
	Climate change adaptation				
Society	Disaster response	Service disruption			•
	Equity and equality				
	Accessibility	Distance travelled		•	•
		Additional time required by public transport			•
	Affordability				
	Acceptability				
	Safety	No. of fatalities	•	•	•
		Fatalities by mode	•		
	Security	People feeling safe			•
	Universal accessibility	Percentage of disable who travel with ease			•
		Percentage of PT i.e. step-free			•
	Quality of life				
	Health	Time spent in active transport			•
	Noise	No. of people exposed to noise			•
	Alternate choice	Modal share	•	•	•
		No. of public transport users			•
Infrastructure	Infrastructure capacity	Volume-capacity ratio	•		
		Capacity of public transport			•
	Smart or intelligent transport				
	System efficiency	Reduce traffic levels			•
	Reliable, mobility and services	Vehicle kilometre	•		
		Delay hours	•		
		Average speed	•		
		Journey time	•		•

Source: (Mayor of London, 2018; New York Metropolitan Transportation Council, 2013; Vancouver City Council, 2012)

transport plan adheres more to the national policy guidelines as compared to the other two cities as it has used higher number of indicators for each of the defined parameters in LTP guideline. There is only one indicator of journey times in the complete indicator set in the plan that is likely to result in selection of strategies for improving travel time by vehicles. The decision maker needs to be cautious by accounting for the impact of car-oriented strategies (like road widening and increasing average speed) on other identified indicators.

Vancouver transport plan includes indicators related to social sustainability only. There is the absence of economic and environmental indicators. Therefore, impact analysis of the proposed actions or strategies on economy and environment is not present in the plan. However, the indicators that have been used in the Vancouver transportation plan are likely to push policies towards achieving social sustainability. As compared to both London and Vancouver, New York transport plan includes more number of indicators that are vehicular mobility centric. Therefore, the planning process is likely to result in selection of strategies that will have negative impact on society and environment.

Different types of indicators of economy, social and environmental sustainability in addition to infrastructure have been used in varying geographical context. The indicators used in the city-level transport plans do not necessarily adhere to the national policies and vision. Number and type of indicators selected within each identified parameters reflect local or regional priorities. Mapping these indicators with sustainability goals can help in understanding if conscious efforts in selecting strategies have been made to achieve the laid goals. In conclusion, application of indicators is limited in the city-level transport plans as compared to the respective national policy document. This finding is similar to the study conducted by (Gudmundsson & Sørensen 2012). There is a strong need to select appropriate indicators in local planning process to achieve national policy vision (Litman 2007; Moussiopoulos et al. 2010; Zachariadis 2005). There is also lack of consensus between local transport actions or plans and national visions and application of indicators can be a strong tool to strengthen the linkage between the two levels. Definition of indicators, measures and data collection procedure in transport planning guidelines issued by national authority can help in addressing the identified gap.

5 Transport Planning Approaches and Indicators in India

In India, different eras of transport planning can be identified. Prior to the issue of NUTP in 2006, transport plan was prepared only as part of the master plan that is to be revised every ten years. The master plan defines existing and future land area required for various activities like residential, commercial and green areas. The allocation of the land is based on the estimated demand for every land use in the horizon year (SPA Delhi 2012). However, the development patterns and infrastructure strategies proposed in the master plan are defined without estimating their impact on societal and the environmental indicators. This had resulted in an increase in inaccessibility, inequity, road fatalities, unaffordability, consumption of resources and the degrading environment in Indian cities (Pucher et al. 2005). Few cities also prepared City Traffic and Transportation Studies (CTTS) for a horizon of 20–25 years. The purpose of CTTS is to estimate travel demand in current and horizon year and provide solutions to meet the estimated demand. The CTTS provided detailed proposal as stated in the master plan. These strategies such as road widening and building flyover therefore focused on increasing the vehicular flows in cities (Malayath and Verma 2013). Detailed project report (DPR) for specific projects were also prepared to provide implementation plan for the identified project that can help in achieving some of the defined targets like reducing travel time.

After the launch of JnNURM scheme, city development plans (CDP) were prepared for all the cities identified under the scheme. CDP looked into multi-sectoral issues specifically pertaining to infrastructural capacities and respective strategies were identified to improve basic infrastructure of the cities. NUTP was adopted in 2006 that mobilized the shift from planning for infrastructure to mobility planning in India (Ministry of Urban Development 2006). JnNURM mandated cities to prepare comprehensive mobility plan to avail funds for improving transport infrastructure from national government such that the transport planning process is aligned with the national policy and targets. CMP provided an approach to integrate land use transport planning and involve transport modelling to provide strategies for enabling better travel demand management in cities. To guide the process, Ministry of Urban Development (MoUD) issued CMP toolkit in 2008 (hereafter referred as toolkit-2008) (ADB 2008). Until the year 2010, 23 of the 65 cities (considered under the JnNURM scheme) had submitted CMP (TERI 2011). The review conducted by TERI in 2011 stated that the CMPs prepared using toolkit-2008 provided investment requirements for the provision of mass rapid transit systems in the cities, without considering the travel behaviour, the travel demand and the impacts of the proposed strategies on accessibility, affordability, safety and environment. Later in 2014, both NUTP and CMP toolkit were revised to lay greater emphasis on low carbon development and smooth adaptation of NUTP at local level (UNEP et al. 2014). The revised CMP toolkit (toolkit-2014) provided a set of indicators to evaluate the existing situation and alternate scenarios and detailed methodology for collecting data. It emphasized on scenario-based approach for planning to achieve the desired targets. Figure 3 shows aims for preparing the four planning documents that have been developed over time in India. Except CDP, all other three documents are core transport infrastructure planning and improvement plans.

As the name and purpose of each of these planning documents suggest, the area considered for analysis differ from each other. DPR is a corridor-oriented approach therefore the scale of the study is smaller than the city. As compared to this, both CTTS and CDP are city-level studies. CMP approach can extend to larger metropolitan area or can also be restricted to city municipal boundaries depending on the jurisdiction type, growth pattern of the city and aim of the local authority.

As defined by the aims and objectives, the approaches or steps undertaken to prepare CDP, CTTS and CMP differs from each other. There is even a difference between the steps or tasks defined in CMP toolkit-2008 and R-toolkit. Figure 4

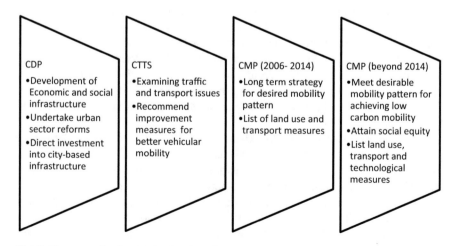

Fig. 3 Transport planning approaches in India

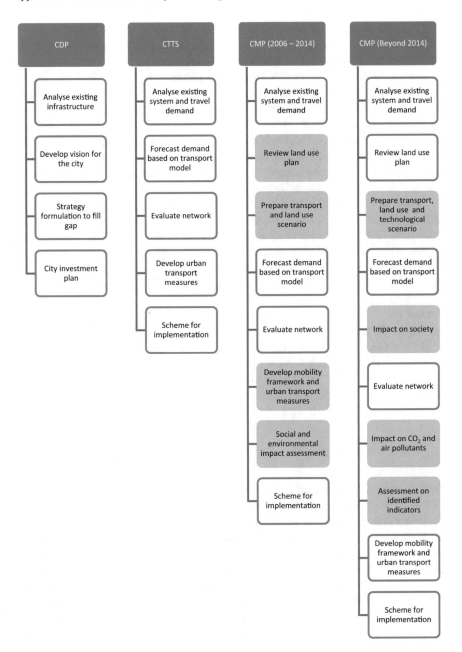

Fig. 4 Tasks undertaken for preparing plans

shows the brief summary of various tasks or steps undertaken for preparing different types of planning documents. In the figure, green shaded cells define the addition or change in the step with regard to the previously proposed approach. For example,

CMP to be prepared beyond 2014 required developing technological scenarios (change in fuel, vehicle technology, etc.) whereas earlier CMP approach looked into development of transport and land use scenarios only.

Aim of the CDP is to provide measures for improving basic infrastructure such that vision for the city with regard to service provision can be achieved. Therefore, it involves evaluation of existing infrastructure, identification of gaps and measures to address the gaps. CTTS is the basic document highlighting transport network improvement strategies. Preparation of CTTS does not involve evaluation of the proposed strategies with regard to social and environmental impacts. The study neither integrates land use transport planning nor considers the impact of changes in the city development pattern or growth pattern on travel demand.

Introduction of CMP helped in including review of land use planning documents and developing land use scenarios to account for existing development patterns and proposals. This therefore enabled integration of land use—transport planning. One of the essential steps included in the CMP approach is to conduct social and environmental impact analysis of the proposed strategies. This therefore would have helped cities in attaining sustainability that was missing in earlier approaches. However, selection of appropriate scenario and strategies was not based on the impact analysis of the alternate scenarios or strategies, therefore resulting in failure of CMPs for achieving the objectives. The new approach defined for preparing CMP addressed this gap. It required evaluating scenarios based on the impacts on identified indicators, such that the goals defined in NUTP can be achieved. As per the R-toolkit, this step needs to be taken up before network improvement, land use and technological improvement strategies are defined.

5.1 Proposed Indicators for Preparing Plans in India

In this section, the type of indicators proposed in CDP, CMP toolkit-2008 and R-toolkit are compared (Table 2). Guidelines for preparing CTTS does not exist therefore it is not included in the analysis. The shaded cells in the vision column highlights that the parameters in the R-toolkit have been addressed as defined in NUTP (2014). Evidently, R-toolkit presents a more comprehensive approach for addressing issues and challenges with regard to attaining sustainable mobility in cities. However, when compared with the goals defined by NUTP 2014, indicators for parameters like economic growth and smart mobility have not been identified in the R-toolkit. The R-toolkit also includes indicators of land use and social heterogeneity. These indicators address the need to integrate land use transport planning and include the dimension of social equity in land use planning. Addressing them can enable increase in accessibility, travel demand management and encourage use of non-motorized transport in the cities. The R-toolkit also discusses the need to include measures of universal accessibility; however, specific indicators and targets have not been defined.

Table 2 Comparison of indicators defined in CDP guidelines, toolkit-2008 and R-toolkit

Agenda or parameters	Vision	Specific indicators	CDP guidelines	Toolkit-2008	R-toolkit
Economy	Cost effective	Cost-benefit ratio		•	
	Finance	Trends in investment			•
		Percentage subsidies granted			•
		Percentage of people owning public transport passes			•
		Cost borne by operators			•
Environment	GHG emissions	Emissions per passenger km			•
		Lifecycle emissions of different modes			•
	Land consumption and waste	Per capita consumption of land for transport activity			•
Society	Equity and equality	All indicators disaggregated by mode			•
	Accessibility	Percentage of Household within 10 min walking distance to public transport			•
		Distance travelled			•
	Urban form	Land use mix			•
		Density			•
		Income level heterogeneity			•
	Inter-modal	Number of interchanges			•
	Affordability	Average cost per trip		•	
		Affordability of public transport			•
	Safety	Fatality rate		•	•
		Risk exposure mode wise			•
		Risk imposed by mode			•
	Security	People feeling safe			•
		Percentage of road lighted			•
		Percentage of area lighted	•		
	Universal accessibility				•
	Health	Percentage of people exposed to air pollution			•
		Percentage of people exposed to noise			•
	Alternate choice	Modal share		•	•
		No. of public transport users			
		Slow moving vehicles share		•	
Infrastructure	Infrastructure capacity	Volume-capacity ratio		•	
		Kernel density of roads and public transport stops			•
		Footpath per unit road length		•	•
		Number of public transport per million people		•	
		Road length	•		
		Number of buses	•		
		Bus capacity per passengers	•		
	Reliable, mobility and services	Vehicle kilometre		•	
		Average speed		•	•
		Journey time		•	•

Table 3 Evaluation of indicators proposed for preparing CMP in toolkit-2008

	Analysis
Evaluate existing traffic and transport situation	Congestion; walkability; city bus transport; safety; IPT; slow moving vehicle; trip distribution; non-motorized vehicle; passenger vehicles
Evaluate alternate scenarios	Potential for developing public transport; total travel time and average speed; volume–capacity ratio; economic indices-net economic cost
Select priority measures	Consistency with policy framework; consistency with strategic transport network framework; impact on congestion; promotion of public transport; traffic safety; cost-effectiveness; constraints

In the toolkit-2008, three different sets of indicators are proposed to analyze the existing situation, select development scenarios and prioritize strategies (Table 3). Therefore, it is difficult to analyze cost and benefits of the proposed strategies with respect to the base year situation. The indicator sets defined in the toolkit are mostly vehicle centric (travel time saving and congestion) and does not include indicators of accessibility. The toolkit proposes measuring overall safety in the city but measuring safety for different modes of transport is not proposed. This, therefore, does not help in identifying the vulnerable road users and desirable interventions to improve safety in the cities. As per the toolkit-2008, institutional and policy framework should play a major role in prioritizing strategies. There are also biases while choosing weights for prioritizing strategies.

Overall, the three indicator sets are not in harmony with each other and lay more emphasis on improving vehicular mobility in toolkit-2008 (mostly personal vehicles). This approach therefore does not helps in achieving the aims of improving mobility of the people. The indicators do not account for evaluation of the scenarios and strategies with regard to the costs and benefits imparted to the society, environment and economy.

As compared to the CMP approach, CDP guideline mentions need to assess transport infrastructure with regard to its availability. Therefore, indicators like road length and bus per passenger are proposed. The indicators related to assessing economic, environmental and social sustainability are not part of the CDP process.

5.2 Application of Indicators in India

To assess the adaptation of R-toolkit two mobility plans of Salem and Udaipur are compared in Table 4. Development of R-toolkit was a part of the project 'promoting low carbon transport in Indian cities' funded by United Nation Environment Programme, a consortium of Denmark Technical University, Indian Institute of Technology Delhi, Indian Institute of Management Ahmedabad and Center for

Table 4 Application of indicators in mobility plan of Udaipur and Salem

Agenda or parameters	Vision	Specific indicators	Udaipur (2014)	Salem (2015)
Economy	Finance	Trends in investment		
		Percentage subsidies granted	✓	
		Percentage of people owning passes for public transport	Not applicable (NA)	
		Cost borne by operators		
Environment	GHG emissions	Emissions per passenger km	✓	✓
		Lifecycle emissions by mode		
	Land consumption and waste	Per capita consumption of land for transport activity		
Society	Equity and equality	All indicators disaggregated by income and gender	✓	
	Accessibility	Percentage of Household within 10 min walking distance to public transport	✓	
		Percentage of area covered by bus routes		✓
		Distance travelled	✓	✓
	Urban form	Land use mix		
		Density		
		Income level heterogeneity		
	Inter-modal	Number of interchanges	✓	
	Affordability	Expenditure by income	✓	✓
		Affordability of public transport	NA	
	Safety	Fatality rate		✓
		Risk exposure mode wise		
		Risk imposed by mode		
	Security	People feeling safe	✓	
		Percentage of road lighted	✓	
	Universal accessibility			
	Health	Percentage of people exposed to air pollution		
	Noise	Perecentage of people exposed to noise		
	Alternate choice	Modal share	✓	✓
		Slow moving vehicles share		✓
Infrastructure	Infrastructure capacity	Footpath per unit road length	✓	✓
		Road length		✓
		Kernel density of roads and public transport stops		
	Reliable, mobility and services	Average speed	✓	✓
		Journey time	✓	

Environmental Planning and Technology Ahmedabad. As part of the project, three case-study cities (Rajkot, Vishakhapatnam and Udaipur) were selected to assess adaptation of the R-toolkit for preparing CMP. CMP of Udaipur was developed by UMTC and it was monitored during development by the UNEP project team. Whereas, mobility plan for Salem was developed in 2015 after the R-toolkit was issued. Since development of CMP for Udaipur was monitored by the UNEP project team developing R-toolkit and CMP for Salem was developed independently, the two CMPs are taken up for reviewing and comparison purposes. The two cities are also selected because their plans were prepared in the similar period and are of small size as per the classification given by Jain and Tiwari (2016).

In Table 4, the shaded cells in specific indicators column present the indicators that have been proposed in the R-toolkit. For both Udaipur and Salem, most of the indicators have been measured except indicators related to safety and exposure of population to air and noise pollution. In CMP of Salem, indicators of security are also not measured.

5.3 Discussion

Since, the three approaches differ from each other and the proposed indicators set also varies, the proposal of strategies using each of the defined processes are likely to vary. CDP approach is likely to result in selection of strategies for expanding infrastructure. Less thrust is likely to be there on proposal of policies to manage travel demand. Toolkit-2008 emphasize on attaining reliable services and infrastructure provision. It also includes indicators of social sustainability like fatality rate and affordability; however, environmental indicators are missing. Therefore, policies and strategies defined in CMP prepared between 2006 and 2014 are likely to be related to infrastructure development for pedestrians, public transport users and car users. How these proposed projects will affect environment are not part of the planning document.

R-toolkit presents a promising approach for defining policies and strategies at local level; however, its adaptation is likely to be limited. This is likely to be attributed to the following reasons –

- Specific indicators related to measuring social heterogeneity, universal accessibility and urban form are not defined in the R-toolkit.
- R-toolkit does not define methods for measuring indicators related to health hazards that can be adopted by local authorities and/or consultants preparing CMPs.
- Some of the indicators like accessibility to public transport and urban form would require data at spatially disaggregate scale. This may be difficult to obtain in majority of the Indian cities (Jain and Tiwari 2017a).

Preparing CMP using R-toolkit requires a multi-disciplinary approach—modelling emissions and noise pollution, quantifying impact on different sections of society and developing travel demand models for all modes. Local authorities may not have the desirable capacity to consider all the aspects. Failure to measure some of the specified indicators can result in the failure of the CMP to achieve the desired objectives. It is therefore recommended to conduct capacity building workshops periodically with the organizations engaged in preparing the CMP. External experts can also be engaged by the organizations, if they do not have the desired capacity. R-toolkit needs to be revised based on the experiences obtained from the CMP prepared after 2014. Some of the revisions may include appropriate indicators for measuring social heterogeneity, urban form and universal accessibility. Specific methods for measuring lifecycle emissions by mode and health hazards need to be developed and included in the R-toolkit. In addition, there is a need to integrate

CMP approach in the formal planning framework and align CMP with the existing development plans like master plan. This would require transformations in the existing institutional and legal framework. After, CMP has been prepared monitoring mechanism to check implementation of the projects proposed in CMP needs to be developed. This will help in quality assessment and increasing assurance that the objectives of NUTP are achieved.

6 Conclusion

Globally, take-up of sustainable transport planning approach has increased that has resulted in increased adoption of scenario-based approach for preparing plans. Countries have identified different priority areas within sustainability agenda and accordingly, the application and choice of indicators vary. Although, national policy guidelines specify use of indicators for preparing transport plans, however limited number of indicators per parameter (defined in national guidelines) exists in the city level transport plans. This therefore is likely to result in delinking strategies defined in transport plans from national vision. Appropriate selection, definition and methodology for measuring indicators in national policy documents can help in strengthening the linkages.

In context of India, there had been a change in the way transport plans have been prepared after 2006. The initially adopted approach of preparing CTTS was limited to demand estimation and formulation of strategies to meet the estimated demand. Development of CMP using either toolkit-2008 or R-toolkit has not only enabled integrated approach of planning but also enabled assessment of strategies and scenarios with regard to indicators. R-toolkit provided further enhancement to the mobility planning approach by incorporating specific indicators of accessibility, affordability, air pollution, safety, security and health. However, since CMP is not a statutory document i.e. it is not legally binding and it is not mandatory for all the cities to develop it, the adoption and application of CMP is therefore limited. Legal and institutional framework needs to be developed for preparing CMP and implementation of the proposed strategies in the CMPs.

Impact of urban transport on society and environment is likely to increase in the near future. This is attributed to the urbanizing population, spatially expanding cities and growing economy. There is therefore a need to identify short and long-term strategies that are specific to the local context to manage travel demand in sustainable way. This requires development of mobility or transport plans at city level, where strategies are evaluated against the identified indicators. It has been identified that although R-toolkit has included an extensive list of indicators; however, methods for measuring some of the essential indicators related to health, security and accessibility for differently abled people have not been defined. There is a need to include specific methods to measure indicators related to health hazard and lifecycle emissions that can be adopted by the local authorities and consultants preparing the CMPs. In addition, appropriate capacity building programs for local authorities need to be conducted to enable development of mobility plans that can help in meeting the laid sustainability goals. Overall, the agenda of sustainable mobility in Indian cities can be achieved by further transforming existing legal and institutional framework, capacity building of local authorities, developing monitoring mechanisms to assess implementation of CMP and revising the issued R-toolkit. This can help in ensuring implementation of nationally laid goals at city level.

References

ADB. (2008). *Module 1: Guidelines for preparing comprehensive mobility plan*. Asian Development Bank, Ministry of Urban Development.

African Union. (2015). *First ten year implementation plan 2014–2023*.

Amekudzi, A. A., Khisty, C. J., & Khayesi. M. (2009). Using the sustainability footprint model to assess development impacts of transportation systems. *Transportation Research Part A: Policy and Practice, 43*(4), 339–348.

Asmaa, A., Brahim, G., & Esteve, A. L. (2012). Transportation planning: A comparison between Moroccan and Spanish decision making process. *Open Transportation Journal, 6*, 1–10.

Bohler-Baedeker, S., Kost, C., & Merforth, M. (2015). *Urban mobility plans—national approaches and local practice*. GIZ, sustainable urban transport technical document# 13.

Boyko, C. T., Gaterell, M. R., Barber. A. R. G., Brown, J., Bryson, J. R., Butler, D., et al. (2012). Benchmarking sustainability in cities: The role of indicators and future scenarios. *Global Environmental Change, 22*(1), 245–254.

Bührmann, S., Wefering, F., & Rupprecht, S. (2011). *Developing and implementing a sustainable urban mobility plan*. Eltisplus Project, Rupprecht Consult.

Castillo, H., & Pitfield, D. E. (2010). ELASTIC—A methodological framework for identifying and selecting sustainable transport indicators. *Transportation Research Part D: Transport and Environment, 15*(4), 179–188.

Commission of the European Communities. (2001). *European transport policy for 2010: Time to decide* Brussels. COM(2001) 370 final.

Commission of the European Communities. (2007). *Sustainable urban transport plans: Preparatory document in relation to the follow-up of the Thematic Strategy on the Urban Environment* Brussels. Technical report—2007/018.

Commission of the European Communities. (2009). *Action plan on Urban Mobility* Brussels. COM(2009) 490 final.

CST. (2002). *Definition and vision of sustainable transportation*. The Centre for Sustainable Transportation.

CST Canada. (2005). *Defining sustainable transportation: Draft 2*. Canada: The Centre for Sustainable Transportation.

Dale, V. H., & Beyeler, S. C. (2001). Challenges in the development and use of ecological indicators. *Ecological Indicators, 1*(1), 3–10.

Department for Transport. (2009). *Guidance on local transport plans*. London: DfT.

DETR. (2000a). *Guidance on full local transport plans*. London: Department of the Environment, Transport and the Regions.

DETR. (2000b). *Local quality of life counts: A handbook for a menu of local indicators of sustainable development*. London: Department of the Environment, Transport and the Regions.

El-Geneidy, A., Diab, E. l., Jacques, C., & Mathez, A. (2013). *Sustainable urban mobility in the Middle East and North Africa Thematic study prepared for global report on Human Settlements 2013*.

Eltis. (2015). Opava: Applying SUMP principles to the planning process (Czech Republic). Retrieved, 2015, from http://www.eltis.org/discover/case-studies/opava-applying-sump-principles-planning-process-czech-republic.

Euromed Transport Forum. (2013). *Regional transport action plan for Mediterranean region (RTAP)—2014–2020.*

European Commission. (2013). *Study to support an impact assessment of the urban mobility package—Activity 31: Sustainable urban mobility plans.*

Federal Highway Administration USA. (2012). *Moving ahead for progress in the 21st century Act: H.R. 4348.*

FHA & FTA. (2015). *The transportation planning process: key issues - a briefing book for transportation decision-makers, officials, and staff.* Federal Highway Administration, Federal Transit Administration.

Gilbert, R., Irwin, N., Hollingworth, B., Blais, P., Lu, H., & Brescacin, N. (2002). *Sustainable transportation performance indicators: Report on phase III.* Canada: Center for Sustainable Transportation.

Gudmundsson, H., & Sørensen, C.H. (2012). Some use - Little influence? On the roles of indicators in European sustainable transport policy. *Ecological Indicators, 35,* 43–51.

Haghshenas, H., & Vaziri, M. (2012). Urban sustainable transportation indicators for global comparison. *Ecological Indicators, 15*(1), 115–121.

Hidalgo, D., & Huizenga, C. (2013). Implementation of sustainable urban transport in Latin America. *Research in Transportation Economics, 40*(1), 66–77.

House of Commons. (2006). *Local transport planning and funding: Twelfth report of session 2005–06.* London: Transport Commitee.

Jain, D., & Tiwari, G. (2016). How the present would have looked like? Impact of non-motorized transport and public transport infrastructure on travel behavior, energy consumption and CO_2 emissions—Delhi, Pune and Patna. *Sustainable Cities and Society, 22,* 1–10.

Jain, D., & Tiwari, G. (2017a). Population disaggregation to capture short trips—Vishakhapatnam, India-. *Computers, Environment and Urban Systems, 62,* 7–18.

Jain, D., & Tiwari, G. (2017b). Sustainable mobility indicators for Indian cities: Selection methodology and application. *Ecological Indicators, 79,* 310–322.

JICA. (2014). The project on integrated urban development master plan for the city of Nairobi in the Republic of Kenya.

Leo, A., Morillon, D., & Silva, R. (2017). Review and analysis of urban mobility strategies in Mexico. *Case Studies on Transport Policy, 5*(2), 299–305.

Litman, T. (2007). Developing indicators for comprehensive and sustainable transport planning. *Transportation Research Record: Journal of the Transportation Research Board,* 2017.

Litman, T. (2009). Sustainable transportation indicators: A recommended research program for developing sustainable transportation indicators and data. *Transportation Research Board Annual Meeting, 2009,* 09–3403.

Lopez-Lambas, M.E., Corazza, M.V., Monzon, A., & Musso, A. (2013). Rebalancing urban mobility: A tale of four cities. Proceedings of the Institution of Civil Engineers: *Urban Design and Planning, 166*(5), 274–287.

Malayath, M., & Verma, A. (2013). Activity based travel demand models as a tool for evaluating sustainable transportation policies. *Research in Transportation Economics, 38*(1), 45–66.

May, A. D., Kelly, C., Shepherd, S., & Jopson, A. (2012). An option generation tool for potential urban transport policy packages. *Transport Policy, 20,* 162–173.

May, A. D., Page, M., & Hull, A. (2008). Developing a set of decision-support tools for sustainable urban transport in the UK. *Transport Policy, 15*(6), 328–340.

Mayor of London. (2018). *Mayor's transport strategy.*

Ministry of Transport New Zealand. (2008). *The New Zealand transport strategy.*

Ministry of Urban Development. (2006). *National urban transport policy.* Government of India.

Ministry of Urban Development. (2014). *National urban transport policy.* Government of India.

Miranda, H. d. F., & Silva, A. N. R. d. (2012). Benchmarking sustainable urban mobility: The case of Curitiba, Brazil. *Transport Policy, 21,* 141–151.

Moussiopoulos, N., Achillas, C., Vlachokostas, C., Spyridi, D., & Nikolaou, K. (2010). Environmental, social and economic information management for the evaluation of sustainability in urban areas: A system of indicators for Thessaloniki, Greece. *Cities, 27*(5), 377–384.

New York Metropolitan Transportation Council. (2013). *Plan 2040: Regional transportation plan —New York.*

Ottawa City Services. (2013). *Transportation master plan Ottawa 2013.*

Pucher, J., Korattyswaropam, N., Mittal, N., & Ittyerah, N. (2005). Urban transport crisis in India. *Transport Policy ,12*(3), 185–198.

Republic of Kenya. (2009). *Integrated national transport policy: Moving a working nation.* Ministry of Transport.

Richardson, B. C. (2005). Sustainable transport: Analysis frameworks. *Journal of Transport Geography, 13*(1), 29–39.

Rosqvist, L. S., & Wennberg, H. (2012). Harmonizing the planning process with the national visions and plans on sustainable transport: The case of Sweden. *Procedia—Social and Behavioral Sciences, 48*, 2374–2384.

Rupprecht Consult. (2011). *The State-of-the-Art of sustainable urban mobility plans in Europe.* ELTISplus Project.

SPA Delhi. (2012). *Alternatives to master plan approach.* Delhi: Department of Urban Planning, School of Planning and Architecture.

TERI. (2011). *Review of comprehensive mobility plans.* Climate Works Foundation.

The President of the Republic. (2012). *The guidelines of the national policy on urban mobility: Law no. 12,587.*

Toth-Szabo, Z., & Várhelyi, A. (2012). Indicator framework for measuring sustainability of transport in the city. *Procedia—Social and Behavioral Sciences, 48*, 2035–2047.

Tyler, N., Ramírez, C., Ortegón, A., Bohórquez, J. A., Suescún, J. P. B., Velásquez, J. M., Perez, M. A., Galarza, D. C., & Peña, A. J. (2013). *Proposal for a national transport strategy for low carbon cities: Colombia 2030.*

UNEP, IIM Ahmedabad, IIT Delhi, & Cept University. (2014). *Comprehensive Mobility Plans (CMP): Preparation toolkit—revised.* Institute of Urban Transport, Ministry of Urban Develoment, Government of India.

UNFCCC. (1997). *Kyoto protocol to the united nations framework convention on climate change.* United Nations.

United Nations Sustainable Development. (1992). *Agenda 21, United Nations conference on environment and development.*

USDOT. (2006). *Strategic plan 2006–2011.* US Department of Transportation (USDOT).

Vancouver City Council. (2012). *Transportation 2040: Plan as adopted by Vancouver city council.*

Vasconcellos, E. A. (2018). Urban transport policies in Brazil: The creation of a discriminatory mobility system. *Journal of Transport Geography, 67*, 85–91.

Zachariadis, T. (2005). Assessing policies towards sustainable transport in Europe: An integrated model. *Energy Policy, 33*, (12) 1509–1525.

Zhou, J. (2012). Sustainable transportation in the US: A review of proposals, policies, and programs since 2000. *Frontiers of Architectural Research, 1*(2),150–165.

Printed in the United States
By Bookmasters